UA Ruhr Studies on Development and Global Governance

Band 71

UA Ruhr Studies on Development and Global Governance

vormals UAMR Studies on Development and Global Governance,
Bochum Studies in International Development und
Bochumer Schriften zur Entwicklungsforschung
und Entwicklungspolitik

Band 71

The UA Ruhr Graduate Centre for Development Studies is a collaboration project
between the Institute of Development Research and Development Policy (IEE),
Ruhr-University Bochum, the Institute of Political Science (IfP) and the Institute for
Development and Peace (INEF), both located at the University Duisburg-Essen.
The Centre is part of the University Alliance Ruhr (UA Ruhr) which aims at establishing the
region as a cluster of excellence in research and training.

Herausgegeben für das UA Ruhr Graduate Centre
for Development Studies von:

Prof. Dr. Matthias Busse, Prof. Dr. Tobias Debiel,
Prof. Dr. Eva Gerharz, Prof. Dr. Christof Hartmann,
Prof. Dr. Markus Kaltenborn, Prof. Dr. Helmut Karl,
Prof. Dr. Wilhelm Löwenstein

Casper Boongaling Agaton

A Real Options Approach to Renewable and Nuclear Energy Investments in the Philippines

Logos Verlag Berlin

λογος

UA Ruhr Studies on Development and Global Governance

Herausgegeben von:

Prof. Dr. Matthias Busse, Prof. Dr. Tobias Debiel,
Prof. Dr. Eva Gerharz, Prof. Dr. Christof Hartmann,
Prof. Dr. Markus Kaltenborn, Prof. Dr. Helmut Karl,
Prof. Dr. Wilhelm Löwenstein

Institut für Entwicklungsforschung und Entwicklungspolitik
Ruhr-Universität Bochum
Universitätsstr. 150
D-44801 Bochum

Telefon: +49(0)234/32-22418
Telefax: +49(0)234/32-14294
E-mail: IEEOffice@ruhr-uni-bochum.de
http://www.uar-graduate-centre.org

Coverbild: © depositphotos.com/Tsyhanova

Bibliographic information published by the Deutsche Nationalbibliothek

The Deutsche Nationalbibliothek lists this publication in the Deutsche Nationalbiblio-
grafie; detailed bibliographic data are available in the Internet at http://dnb.d-nb.de.

ISBN 978-3-8325-4938-1
ISSN 2363-8869

Logos Verlag Berlin GmbH
Comeniushof, Gubener Str. 47,
10243 Berlin
Tel.: +49 (0)30 / 42 85 10 90
Fax: +49 (0)30 / 42 85 10 92
http://www.logos-verlag.de

ACKNOWLEDGMENTS

I would like to express my sincerest gratitude to my thesis supervisor Prof. Dr. Helmut Karl for the continuous support and guidance for my PhD study and research. I also would like to thank my second supervisor, Apl. Prof. Dr. Nicola Werbeck, for your guidance and considerate evaluation, and Prof. Dr. Wilhelm Löwenstein as the chairperson in my oral examination. I would like to extend my gratitude to Prof. Graham Weale for giving significant inputs and suggestions to improve my work.

I am grateful to RUB, particularly to IEE, for their support. I am indebted to German Academic Exchange Service (DAAD) - Graduate School Scholarship Program whose financial support gave me a rare opportunity to pursue my PhD studies in Germany. I acknowledge the German Research Foundation (DFG) Open Access Publication Fund of the RUB for the support in publishing my research articles. I also acknowledge IEE and RUB Research School for providing enough seminars and workshops that broaden my knowledge and skills in development research and publishing. I thank the former and current PhD coordinators Dr. Anja Zorob, Dr. Martina Shakya, and Dr. Gabriele Baecker, as well as the staff and student assistants of IEE, for the administrative and technical support during my stay in Bochum. I offer special thanks to my colleagues Lesley Hope, Farah Asna Ashari, Mabel Hoedoafia, Maruf Lutfur, Mohamed Dawude Temory, Zena Mouawad, Britta Holzberg, and the rest of cluster 3 members for fruitful discussions, suggestions, and moral support.

I am greatful to the staff of the Philippines' Department of Energy and National Economic and Development Authority for providing data and suggestions to improve my research. Special appreciation is given to my online advisors Exequiel Cabanda, Jean Centeno, Madeleine Vinuya-Cinco, Zemma Ardaniel, Fatima Del Prado, Liezl

Basilio, and Iamel Montoya for valuable insights and free consultations whenever I face difficulties in understanding my work. I will forever be thankful to Dr. Jennifer Dela Torre, Dr. Hiroaki Miyamoto, Dr. Makoto Kakinaka, Dr. Corazon Morilla, and Dr. Victorina Acero for your invaluable support and recommendations for my PhD applications. Special thanks to Dr. Koji Kotani for introducing this research, providing research materials, and sharing your ideas and experties during my masters, which served as the foundation for this research. I extend my greatest appreciation to the Filipino communities in Germany for providing moral, spiritual, and financial support during my stay in Germany. *Maraming salamat sa inyo mga kababayan! Mabuhay po kayo!* To Dr. jur. Kurt Conscience for offerring your abode to be my home for four years. And to all the people I met around the world throughout this journey, for sharing your expertise, experiences, resources, life lessons, and wisdom.

Lastly, I thank my family and friends for unconditional love and continued support that propel me forward.

Bochum, August 2018

Casper Boongaling Agaton

CONTENTS

iii

LIST OF TABLES

LIST OF FIGURES

LIST OF ABBREVIATIONS

ADF	Augmented Dickey-Fuller
BAU	business as usual
BNEF	Bloomberg New Energy Finance
CER	certied emission reduction
CSP	concentrated solar power
DCF	discounted cash ow
DOE	Department of Energy
EEA	European Environmental Agency
EIA	Energy Information Administration
ENPV	Expected net present value
EU	European Union
FiT	feed-in tariff
FS-UNEP	Frankfurt SchoolUnited Nations Environment Program
GBM	Geometric Brownian motion
IAEA	International Atomic Energy Agency
IEA	International Energy Agency
IEC	International Energy Consultants
IRENA	International Renewable Energy Agency
IRR	internal rate of return
LSMC	Least Squares Monte Carlo
NEA	Nuclear Energy Agency
NPV	net present value
OECD	Organization for Economic Cooperation and Development
O&M	operations and maintenance
PALECO	Palawan Electric Cooperative
PEWMA	Poisson Exponentially Weighted Moving Average
PSA	Philippine Statistics Authority
PTC	production tax credit
PV	photovoltaic
R&D	research and development
RE	renewable energy
REC	renewable energy credit
REN21	Renewable Energy Network Policy for the 21st Century
RES	renewable energy resources
ROA	Real Options Approach
ROI	return on investment
WREC	World Renewable Energy Congress

Chapter 1

Introduction

In order to reduce the risk of climate change and support a sustainable future, governments and businesses around the world are investing in clean energy. In the recent years, the costs of clean energy, particularly renewable energy, are declining fast and becoming cost-competitive against fossil fuel-based alternatives in many countries (OECD (2016)). These boost the growth of new clean energy investments setting a record high in 2015 of US$249 billion, with a shift in geographic concentration in the Asia-Pacific region (see Figure 1.1). In 2016, investment in renewable energy (excluding large hydro) accounts to US$227 billion bringing additional 138.5GW of global power capacity (BNEF (2017a); FS-UNEP (2017)). To date, climate change policies and improving cost-competitiveness are the main drivers of investment in renewable energy to sustain its continuous growth of share of world electricity generation (FS-UNEP (2017)). Despite of this, there are still obstacles that impede investments in renewable energy. These include national monopolies in developing countries that are not so familiar with variable wind and solar generation; balancing variable generation and storage; lack of investor confidence due to political events or energy policy; lacking of financial option for some developing countries; and policy obstacles including trade, partnership, electricity market, financial market and investment policies; (FS-UNEP (2017); OECD (2016)). In the Philippines, the strong economic growth and increasing energy demand have caused pressure on the country's power sector to invest in more sustainable sources of energy. At present, fossil-based sources still dominate the power sector with 45% of total energy mix generated from coal in 2015 (see Figures 1.2 and

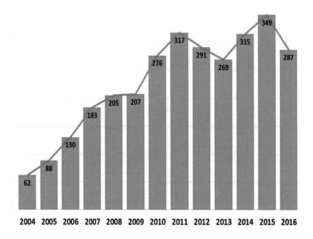

Figure 1.1: World total annual new investment in clean energy in US$bn.

source: BNEF (2017a)

1.3). Despite the natural reserves of fossil fuels, the country is heavily dependent on imported coal (75% of mix) from Indonesia, China, and Australia, and oil (87%) from Saudi, Kuwait, UAE, and others (DOE (2016a)). With the volatile prices of fossil fuels in the world market, sudden changes in price may eventually affect the country's energy supply and security. Further, the recent focus on anthropogenic climate change in the global policy agenda highlights the negative consequences of using fossil fuels which combustion releases greenhouse gases. The increasing energy demand along with concerns on its limited supply, decommissioning old power plants, fuel price volatility, national security problems, and negative environmental effects of using fossil fuels serve as the impetus for finding alternative sources to meet the country's energy needs.

To address the country's problem of energy security, the government started its nuclear energy program in 1973 during the world oil crisis. After careful evaluation of proposed project, the construction of Bataan Nuclear Power Plant began in 1976. The

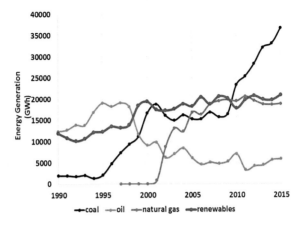

Figure 1.2: Power Generation in the Philippines by source.

source: DOE (2016a)

plant that costed US$2.3 billion was nearly finished in 1984 and designed to supply 621MW of electricity (Magno, 1998). However, due to numerous protests related to nuclear disasters (Three Mile island accident in 1979 and Chernobyl accident in 1986), controversies of corruption, potential threat to public health due to its earthquake zone location, and nuclear safety, the succeeding administration discontinued the program (Beaver (1994), Lee and So (1999)). In the recent years, the government is considering to rehabilitate the mothballed plant and construct four additional nuclear power plants as a long-term option for energy source in the country (IAEA (2016)).

Another promising alternative to meet the country's energy demand are renewable energies(RE). They are expanding both in terms of investment projects and in terms of geographical spread across the country. In 2015, renewable energy accounts to 25% of total energy generation mix (see Figure 1.3). The country is aiming to increase its current capacity to 60% by 2030 by developing localized renewable energy resources

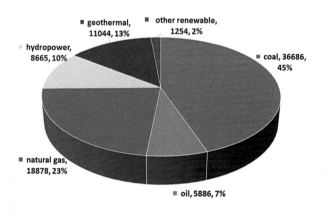

Figure 1.3: Power Generation in 2015 by source (in GWh).

source: DOE (2016a)

(DOE (2012)). With a promise of cleaner and more sustainable source of energy,
investment in renewable energies in the Philippines is still facing various challenges.
These include the competitive prices of imported fossil fuels, political issues ensuring
the continuity of the project, risks of natural calamities, and uncertainties with regard
to variability of renewable sources. These give an impetus to make a study that suggests
energy investment strategies that address the country's problem on energy security
and sustainability.

This study proposes a general framework of investment decision-making for shifting
technologies from fossil fuels to various alternative energy sources that can be applied
to developing economies, particularly to fossil fuel-importing countries. By taking the
case of the Philippines, this dissertation applies the real options appoach (ROA) to
analyze various investment scenarios. Traditionally, valuation of public investments
in the country include net present value, internal rate of return, discounted cash flow,

return on investment, and others. However, these methods do not cover highly volatile and uncertain investments as they assume a definite cash flow. This assumption makes these methods underestimate the investment opportunities leading to poor policy and decision-making process, particularly to energy generation projects. These approaches do not allow investors to define the optimal timing to invest and estimate the value of project uncertainties(Santos et al. (2014)). Real options approach (ROA) overcomes these limitations by combining uncertainty and risk with flexibility of investment as potential factors that give additional value to the project (Brach (2003)). Compare with other approaches, ROA combines three important characteristics of investment decision such as the uncertainty of the future cash flows from investment; irreversibility of investment; and the flexibility in the timing of investment opportunity (Baecker (2007)). The overview of ROA along with other methods used to evaluate energy investments is summarized in Table 1.1.

Table 1.1: Summary of valuation methods used in energy investment projects.

Method	Definition	Numerical Solution	Decision Criterion to Implement the Project	Applications
Net Present Value (NPV)	Sum of the present value of all cash flows produced by the project	$NPV = \sum_{t=1}^{n} \frac{NR_t}{(1+i)^t} - \sum_{t=0}^{n-1} \frac{I_t}{(1+i)^t}$	$NPV > 0$	Oil and Gas industries. Renewable energy investments/projects.
Internal Rate of Return (IRR)	Represents the discount rate that equalizes the NPV to zero.	$0 = \sum_{t=1}^{n} \frac{NR_t}{(1+IRR)^t} - \sum_{t=0}^{n-1} \frac{I_t}{(1+IRR)^t}$	$IRR > k$	Renewable energy investments/projects.
Return on investment (ROI)	Measures the relation between the present value of cash flows and the necessary investments to implement the project.	$ROI = \frac{\sum_{t=1}^{n} \frac{NR_t}{(1+i)^t}}{\sum_{t=0}^{n-1} \frac{I_t}{(1+i)^t}}$	$ROI > 1 \ (NPV > 0)$	Renewable energy investments/projects.
Payback period	Period of time required to recover the investments.	$P = \frac{\sum_{t=0}^{n-1} \frac{I_t}{(1+i)^t}}{\frac{NR_t}{(1+i)^t}/n}$	$P < n$	Renewable energy investments/projects.
Benefit-Cost Ratio (BCR)	Identify, quantify and weigh the benefits and costs of the investment projects.	$B/C = \frac{\sum_t \frac{R_t - C_t}{(1+i)^t}}{\sum_t \frac{I_t}{(1+i)^t}}$	$\frac{B}{C} > 1$	Renewable energy investments/projects.
Levelized costs (LC)	Compare the energy generation technologies with different characteristics and lifetimes.	$LC = \frac{C_I + C_{O\&M} + C_c + C_d}{E_{act}}$	lowest levelized cost	Energy investments/projects. Energy Market. Power Generation.
Real Options (ROA)	Reformulate the NPV so that the scenarios of great uncertainty, which compose the investments, are considered.	$NPV_{expanded} = NPV_{traditional} + V_{managemen\ t\ flexibility}$	$NPV_{expanded} > 0$	Oil and Gas industries. Renewable energy investments/projects. Energy market. Power generation.

source: Santos et al. (2014)

Note: I_t: investment cash-flow in period t; NR_t: net revenue in period t; i: discount rate; k: reference interest rate or the opportunity cost of capital; n: number of years; R_t-C_t: operation cash-flow in period t; C_I: present investment cost; $C_{O\&M}$: present value of operations and maintenance costs; C_c: present value of fuel costs; C_d: present value of various annual costs; E_{act}: present commulative value of energy production

Recent studies use ROA to valuate investments projects in renewable energy (wind, hydropower, solar PV) and nuclear energy. They employ various uncertainties including electricity demand, public acceptance in nuclear, REC price, cost of delay, FiT, CER, O&M cost, energy production, investment cost, CO2 price, R&D cost, tax credit, subsidy, regulatory policies, non-RE price, and exchange rate. A list of recent ROA literature on various energy sources and investment uncertainties reviewed in this dissertation is summarized in Table 1.2. This dissertation aims to contribute to the existing literature by providing a ROA framework for electricity generation switch from fossil-based to alternative sources of energy including wind, solar, hydropower, geothermal, and nuclear. This further examines how various factors affect energy investment decisions such as multi-period investment model, uncertainty in fossil fuel prices, risk of possible nuclear accident, growth of RE investment, variability in social discount rates and electricity prices, RE investment costs, and negative externality for using fossil fuels. Specific contributions of this study are introduced by chapter in the next subsection.

The ROA in this dissertation presents a model of an investor that is given a specific period to decide whether to invest in alternative energy or to continue using fossil fuel for power generation. After such period, an investor has no other option but to continue using fossil fuel-based energy until the end of the lifetime operation of such technology. The energy switching decision is evaluated in yearly basis using dynamic programming by estimating the option values and trigger price of fossil fuels. Option values are calculated by maximizing the value of either investing in alternative energy or using fossil fuel-based energy for each period of investment. The trigger price is characterized by the minimum price fossil fuel where option values for initial period of investment equals the terminal period. At this price, the value of option to wait, as

Table 1.2: Summary of ROA literature

Author(s) (Year)	Energy source	Uncertainty
Cardin et al. (2017)	nuclear	electricity demand, public acceptance
Eissa and Tian (2017)	solar	REC price, cost of delay
Kim et al. (2017)	hydro	FiT, CER, O& M cost, energy production
Kitzing et al. (2017)	wind	FiT, feed-in premium, green certificates
Loncar et al. (2017)	wind	compound ROA
Tian et al. (2017)	solar	investment cost
Zhang et al. (2017)		electricity price, CO2 price, investment cost
Barrera et al. (2016)	solar	R& D cost
Eryilmaz and Homans (2016)	wind	Tax credit, REC
Fleten et al. (2016)		electricity price, subsidy
Ritzenhofen and Spinler (2016)	wind	FiT adjustment
Sisodia et al. (2016)	wind	regulatory policy changes
Tian et al. (2016)	nuclear	carbon market
Wesseh-Jr. and Lin (2016)	wind	FiT
Zhang et al. (2016)	solar	non-RE cost, FiT, investment cost
Wesseh-Jr. and Lin (2015)	hybrid	non-RE price, R&D funding
Weibel and Madlener (2015)	hybrid	Energy production, FiT, investment cost
Jeon et al. (2015)	hydro	FiT, energy production, interest rate, risk free rate, exchange rate
Kim et al. (2014)	wind	non-RE cost
Abadie and Chamorro (2014)	wind	FiT, energy production, subsidy
Detert and Kotani (2013)	hybrid	non-RE cost
Shi and Song (2013)	nuclear	risk
Rothwell (2006)	nuclear	CER price

characterized by the difference of option values between initial and terminal period of investments, become zero. The optimal timing strategy for switching technology is characterized by the trigger price of fossil fuels that maximize the values of investment in alternative energy. The model is then applied to analyze switching decisions from various fossil fuels (coal and diesel) to different alternative energy sources including nuclear and renewable (wind, solar PV, geothermal, hydro).

1.1 Content of the Thesis

This dissertation [1] is composed of four papers. Chapter 2 focuses on investment with various types of renewable energy sources, chapter 3 compares investments between renewable and nuclear energy, while chapter 4 applies the proposed model with the case of Palawan. Finally, chapter 5 focuses on the drivers that make RE a better option than fossil fuels.

In particular, chapters 2 to 5 of this dissertation deal empirically with the following questions:

- Ch. 2: Is investment in renewable energy sources a better option than continue using coal?

- Ch. 3: Is renewable or nuclear energy a better alternative to coal for electricity generation?

- Ch. 4: Is renewable energy a better option than continue using bunker-fired diesel power plant for electricity generation?

---

[1] Agaton (2018b) available at `https://doi.org/10.13154/294-6119`

- Ch. 5: How investment drivers affect shifting energy source from coal to renewable?

Chapter 2 (published as a chapter entitled "*To Import Coal or Invest in Renewables? A Real Options Approach to Energy Investments in the Philippines*" in the book *Transition Towards 100% Renewable Energy - Selected Papers from the World Renewable Energy Congress WREC 2017* and an article entitled: "*Use coal or invest in renewables: a real options analysis of energy investments in the Philippines*" published in the *Renewables: Wind, Water, and Solar* journal) deals with investment with various renewable energy sources including wind, solar PV, geothermal, and hydropower. The aim of this chapter is to evaluate the comparative attractiveness of either investing on renewable energy sources or continue using coal for electricity generation by considering uncertainty in coal prices. ROA is used to empirically evaluate the investment option value, value of waiting to invest in RE, and optimal timing of switching technologies from coal to renewable sources. A sensitivity analysis is conducted to investigate the dynamics of optimal investment decisions with respect to uncertainty in coal prices and discount rates. Higher uncertainty describes a situation with high volatility in coal prices, while lower uncertainty indicates a more deterministic trend in coal prices.

Chapter 3 (an article entitled "*Coal, Renewable, or Nuclear? A Real Options Approach to Energy Investments in the Philippines*" published in the *International Journal of Sustainable Energy and Environmental Research*) examines investments in renewable energy and nuclear energy. The aim of this chapter is to evaluate the comparative attractiveness of either investing in alternative (renewable or nuclear) energy or continuing the use of coal for electricity generation in the Philippines. ROA under coal price uncertainty is used to analyze investment values and optimal timing of switching technologies from coal to renewable or nuclear energy. This chapter examines how risk of nuclear accident and negative externality affect investment decisions. The nuclear accident scenario describes a situation where nuclear disaster may happen once, at most, in the lifetime of nuclear energy generation. The energy generation terminates once the accident occurs, hence, accident cannot be repeated. Once it happens, the energy generation stops, the plant is decommissioned, and the producer pays the necessary accident costs. Negative externality scenario, on the other hand, describes a situation where government impose externality tax for using coal, renewable, and nuclear. In this scenario, the externality cost for renewable and nuclear are set to a fixed level but varied for using coal. The main reason for this is the hypothesis that changes in externality cost for using coal affect investment decisions for shifting to alternative sources of energy.

Chapter 4 (published as an article entitled "*A real options approach to renewable electricity generation in the Philippines*" in the *Energy, Sustainability and Society* journal) deals with the application of the proposed real options model in the case of Palawan, an island-province run by a bunker-fired diesel power plant for electricity generation. While previous chapters deal with technology switch from coal to alternative energy, this chapter evaluates the comparative attractiveness of either investing in solar PV or continuing to use diesel fuel for electricity generation. This chapter analyzes how the timing of investment in solar PV depends on local electricity price and negative externality for using diesel. Electricity prices varies across the province from various power producers, and are constantly changing due to volatile prices of diesel. The electricity price scenario describes how policy of imposing electricity price ceiling or price floor affects the investment decision of introducing renewable energy into the island. Meanwhile, the negative externality accounts to the health and environment problems associated with combustion of diesel. In this scenario, energy generation from solar PV is assumed to produce minimal or no externality. Thus, this scenario accounts for the effect of varying externality cost of using diesel on switching technologies to renewable energy.

Chapter 5 (an article entitled "Real Options Analysis of Renewable Energy Investment Scenarios in the Philippines" published in Renewable Energy and Sustainable Development journal) analyzes various investment scenarios that make renewable energy a better option than continuing to use coal for electricity generation. Contrary with previous articles, this paper proposes a multi-period investment. This aims to depict the country's situation of increasing energy demand due to rapid economic development and industrialization, as well as the old coal plants closing in various periods. Further, full system switch from coal to renewable, as described in previous papers, is less plausible in reality particularly in developing countries like the Philippines, hence, successive investment is proposed. Applying the ROA, this chapter compares the attractiveness of renewable energy over coal under various investment scenarios in the Philippines. The business-as-usual scenario describes the current energy investment scenario in the country with average 2% growth rate in renewable energy investment which is much lower than the growth rate of energy demand, no GHG prices, high investment costs for RE, high electricity costs, no technological innovations, and highly dependent on imported coal. The next scenario describes an accelerated growth of investment in renewable energy sources. While the country is aiming to increase the current share of energy generation from renewables from 25% to 60% by 2030 at 4% annual growth rate, this goal seems unattainable as the country's electricity demand is increasing at a faster rate than renewable investments. This scenario examines how changing the rate of growth in renewable energy investment affects the option values and trigger prices. The investment cost scenario describes how the decline in overnight cost affect investment in renewables. In the recent years, growth in renewable energy investments is driven by several factors including the improving cost-competitiveness of renewable technologies, policy initiatives, better access to financing, growing demand for energy, and energy security and environmental concerns This scenario focuses on the effect of renewable energy cost on investment option values and trigger prices of coal for shifting technologies. In the electricity price scenario, the effect of changing electricity prices from renewables on option values and trigger prices is analyzed. By changing the value broadly, this scenario presents how potential government actions regarding electricity prices affect investment conditions in renewable energy. The last scenario discusses the effect of carbon prices for electricity generation from coal. Currently, there is no carbon prices in the Philippines. This is study evaluates the effect of imposing carbon tax as proposed in the literatures.

Finally, chapter 6 concludes the commulative dissertation and summarizes the research findings from the published articles. This also presents the study's environmental and energy policy implications, limitations, and recommendations for further research.

Chapter 2

Use coal or invest in renewables: a real options analysis of energy investments in the Philippines

Abstract - This chapter is based on a paper published as a chapter in the book: *Transition Towards 100% Renewable Energy - Selected Papers from the World Renewable Energy Congress WREC 2017*[2] and an article published in *Renewables: Wind, Water, and Solar*[3] journal. The aim of this chapter is to analyze the comparative attractiveness of either investing in various renewable energy sources or continue using coal for electricity generation in the Philippines. Using real options approach, this chapter evaluates the investment value and trigger prices of coal for switching technologies with some scenarios in coal price uncertainty and social discount rate. The results find that investing in renewable energy is a better option than continue using coal for electricity generation. Among renewable energy sources, geothermal is the most attractive to invest to, followed by wind, hydroelectric, and solar photovoltaic.

---

[2]Agaton (2018a)available at https://doi.org/10.1007/978-3-319-69844-1_1
[3]Agaton (2018c)available at https://doi.org/10.1186/s40807-018-0047-2 licensed under *Creative Commons Attribution License 4.0*

2.1   Background

To reduce the risk of climate change and support a sustainable future, governments and businesses around the world are investing in renewable sources of energy such as wind, hydropower, solar, biomass, and geothermal. In the recent years, the costs of renewable energy are declining fast and becoming cost-competitive against fossil fuel-based alternatives in many countries. These boost the growth of renewable energy investments setting a record high in 2015 of US$249 billion, with a shift in geographic concentration in the developing countries around the Asia-Pacic region (BNEF (2016)). In the Philippines, renewable energy accounts to 24% of the total electricity generation in 2016 (Department of Energy DOE (2017)). The country is aiming to increase this value to 60% by 2030 by developing localized renewable energy resources (DOE (2012)). The country's geographic location in the Pacific makes it a good potential for renewable energy generation with 76.6GW wind, 10GW hydropower, 5kWh/m2/day solar, 500MW biomass, 170GW ocean, and 4GW geothermal, (International Renewable Energy Agency IRENA (2017)). Despite the renewable energy potential in the country, investments in these sources are challenged by high startup and technology cost, competitive prices of fossil fuels, and non-viable markets.

This paper proposes a general energy investment framework that can be applied to developing countries. The main objective is to analyze the comparative attractiveness of either continue using coal or switching to renewable energy sources for electricity generation using the case of the Philippines. Applying the real options approach (ROA) under uncertainty, this research evaluates the value of investment and identifies the optimal timing of investment in various types of renewables. A sensitivity analysis is conducted to investigate the dynamics of investment value and optimal timing of investment under the changes in discount rates and volatility of coal prices.

The application of ROA approach is becoming more popular in valuating energy projects as it covers essential characteristics of investment. First characteristic is the irreversibility in which the investment cost cannot be recovered once it is installed (Pindyck (1993)). Second, ROA addresses the uncertainties in investment including interest rates, technological progress, energy policy, and market prices (Kumbaroglu et al. (2008)). Third, is the flexibility in which investors can invest immediately or delay the decision into a lesser risk and more profitable period of investment (Yang et al. (2008)). Recent applications of ROA, particularly in renewable energy investments, include Eissa and Tian (2017) who investigated the real options framework for solar power project considering the renewable certificate price and cost of delay between establishing and operating the solar power plant; Kim et al. (2017) who assessed the renewable energy investment in developing countries with a case study involving a hydropower project in Indonesia; Kitzing et al. (2017) who evaluated offshore wind energy investments in Baltic Sea under uncertainties in feed-in tariffs (FiT), feed-in premiums, and tradable green certificates; Loncar et al. (2017) who used a compound real options valuation method to examine a potential onshore wind farm project in Serbia; and Zhang et al. (2017) on estimating the optimal subsidy for renewable energy power generation project in China by using stochastic process to describe the market price of electricity, CO2 price, and investment cost. Further, Barrera et al. (2016) analyzed the impact of public research and development (R& D) financing on renewable energy projects, specifically on concentrated solar power; Eryilmaz and Homans (2016) examined the investment decisions of US renewable energy producers considering the uncertainties in federal governments continuation of the production tax credit policy and the market prices for renewable electricity credits; Fleten et al. (2016) studied whether investors in renewable energy projects in Norway exert discretion about the timing of investment decisions when they face uncertainties in electricity price and subsidy; Ritzenhofen and Spinler (2016) assessed the impact of adjustments in FiT schemes on investment in renewable energy sources; Sisodia et al. (2016) evaluated the investment strategies in wind-generated energy projects in Portugal under the risk in regulatory changes in Spain; and Wesseh-Jr. and Lin (2016) evaluated whether the feed-in-tariffs outweigh the cost of wind energy projects in China. To the best of authors knowledge, there has not been any study analyzing investment decisions with various renewable energy options applying the ROA under uncertainty. This study contributes to these literatures by applying ROA, to analyze investment strategies of shifting technologies from coal to renewable energy sources (RES) including wind, solar PV, hydropower, and geothermal. Applying the case of the Philippines, this further identifies scenarios where investment in renewables become better alternative to coal for electricity generation.

2.2   Methodology

This study uses the real options approach to identify the trigger prices of coal for switching technologies from coal to renewable energy sources. A series of processes are employed including the evaluation of the net present values (NPV) of investment in renewables and coal, application of stochastic process and Monte Carlo simulations to estimate the expected net present value of using coal, and dynamic optimization to calculate the real option values of either investing in renewables of continue using coal. The ROA framework in this study takes the point of view of an investor who decides to invest in RES within a specific period of investment. Within this period, the investor has the option to delay the investment and select the optimal timing to maximize the project value under uncertainty (Zhang et al. (2017); Hach and Spinler (2016); Pringles et al. (2015)).

2.2.1 Net present value of investment in renewable energy

Adopting the social revenue function described by Detert and Kotani (2013), Twidell and Weir (2015), and Savino et al. (2017), the NPV for shifting to renewable energy sources is described by

$$NPV_R = \sum_{t=0}^{T_R} PV_{R,t} - I_R = \sum_{t=0}^{T_R} \rho^t \pi_{R,t} - I_R \tag{2.1}$$

$$\pi_{R,t} = P_E Q_E - C_R \tag{2.2}$$

where $\pi_{R,t}$ is the annual revenue for after making the technological switch; R is the renewable source including wind, solar, hydro, and geothermal; I_R is the overnight cost for renewables; P_E is the domestic electricity price in the Philippines; Q_E is the annual electricity generated from renewables; C_R is the annual O&M cost; ρ is the discount factor equal to $\frac{1}{1+r}$; r is the social discount rate; t is the period of investment; and T_R is the lifetime of renewable energy generation.

2.2.2 Net present value of using coal

The social revenue for continue using coal for electricity generation is described by

$$NPV_C = \sum_{t=0}^{T_C} PV_{C,t} = \sum_{t=0}^{T_C} \rho^t \pi_{C,t} \tag{2.3}$$

$$\pi_{C,t} = P_E Q_E - P_{C,t} Q_C - C_C \tag{2.4}$$

where $\pi_{C,t}$ is the annual revenue for continuing the use of coal; C_C is the annual O&M cost for coal; Q_C is the amount of coal needed to generate Q_E, T_C is the number of

years the coal can be used after the terminal period; $P_{C,t}$ is the stochastic price of coal
which is described in the next subsection.

2.2.3   Stochastic process and Monte Carlo simulation

In line with previous literatures, this study assumes that price of coal is stochastic
that changes randomly over time and follows Geometric Brownian motion (GBM) with
a drift (Tietjen et al. (2016); Wang and Du (2016); Xian et al. (2015)). Dixit and
Pindyck (1994) presents the stochastic price process as

$$\frac{dP}{P} = \alpha dt + \sigma dz \tag{2.5}$$

where α and σ are parameters of drift and variance representing mean and volatility
of the price process, dt is the infinitesimal time increment, and dz is the increment
of the Wiener process equal to $\varepsilon_t \sqrt{dt}$ such that $\varepsilon_t \ N(0,1)$. Adopting Insley (2002), α
and σ can be approximated using the augmented Dickey-Fuller (ADF) unit root test
for the time series of coal prices. Using 1986-2016 coal prices data, the ADF test in
Table 2.5 implies that the null hypothesis that p_t has a unit root cannot be rejected at
all significant levels. Therefore, coal prices conform with GBM. The estimated GBM
parameters $\alpha = 0.35094$ and $\sigma = 0.22576$ are then employed in generating stochastic
prices of coal using the equation

$$P_{C,t} = P_{C,t-1} + \alpha P_{C,t-1} + \sigma P_{C,t-1} \varepsilon_{t-1} \tag{2.6}$$

where $P_{C,t}$ and $P_{C,t-1}$ are the stochastic prices of coal at periods t and $t-1$; and ε_t is
standard normally distributed such that $\varepsilon_t \ N(0,1)$.

The expected net present value of using coal is calculated using Monte Carlo simulation as described by the equation

$$\mathbb{E}\left\{NPV_{C,j} \mid P_{C,0}\right\} \approx \frac{1}{J}\sum_{j=1}^{J}NPV_{C,j} \approx \mathbb{E}\left\{NPV_C \mid P_{C,0}\right\} \tag{2.7}$$

In this process, the NPV_C is calculated repeatedly in an approximately large number of J times considering the stochastic prices of coal. Computed NPV_C are then averaged to estimate the expected net present value of using coal.

2.2.4   Dynamic optimization and optimal trigger prices

The real options model in this study is described by an investor who is given a period to decide to switch to renewable energy sources. After such period, there is no other choice but to continue using coal until the end of its lifetime. The decision-making process is done annually by maximizing the net present values of each option (coal or renewable). Adopting the work of Detert and Kotani (2013), option value at each period of investment is calculated using dynamic optimization as described by

$$\max_{0 \leq \tau < T+1}\mathbb{E}\left\{\left[\sum_{0 \leq t < \tau}\rho^t\pi_{C,t} + \rho^T NPV_{C,t}\left(1 - \mathbb{I}_{\{\tau \leq T\}}\right) \mid P_{C,0}\right] + \left\{NPV_R\right\}\left(\mathbb{I}_{\{\tau \leq T\}}\right)\right\} \tag{2.8}$$

$$V_t(P_{C,t}) = max\left\{NPV_{R,\Pi_{C,t}} + \rho\mathbb{E}(V_{t+1}(P_{C,t+1}) \mid P_{C,t})\right\} \tag{2.9}$$

where $V_t(P_{C,t})$ is the option value at each price of coal ($P_{C,t}$, T is the length of time where an investor has an option to switch to renewable energy; τ is the period where the switching is made; and $I_{\tau \leq T}$ is an indicator function equal to 1 with $\tau \leq T$, otherwise 0. In the given equation, the investor's problem is to find the optimal timing τ that maximizes the expected NPV of social revenues at each price of coal for every investment period. Estimated option values are plotted in graphs to identify the trigger prices of coal for switching technologies as represented by

$$\widehat{P_C} = inf\left\{P_{C,t} \mid V_0(P_{C,t}) = V_{T_A}(P_{C,t})\right\} \tag{2.10}$$

where $\widehat{P_C}$ is the trigger price or the minimum coal price $P_{C,t}$, while $V_0(P_{C,0})$ and $V_{T_R}(P_{C,0})$ are the maximized values of the investment at time $t = 0$ and $t = T_R$ (Dixit and Pindyck (1994)); and Davis and Cairns (2012))

2.2.5   Data and Scenarios

To estimate the parameters for the optimization problem, the data are gathered from DOE. A standard quantity of electricity generation Q_E is set to 2165 GWh such that all renewable energy sources produce the same average annual output. This amount proposes 5% of the energy generation from coal to be replaced by renewable sources. The investment costs and other costs associated with the generation of electricity from various sources are then estimated. All parameters used in this study are described in Appendix Table 2.4.

To describe several investment environments in the Philippines, this study analyses the sensitivity of optimal investment decisions with respect to uncertainty in coal prices and discount rates. Higher uncertainty describes a situation with high volatility in coal prices. Meanwhile, lower uncertainty indicates a more deterministic trend in coal prices. Several discount rates are also employed from the present 10%, to 12.5%, 7.5% and 5%.

2.3   Results and Discussion

The dynamic optimization of the real options model in this study yields two main results: option values and optimal trigger price. The option values at every period of investment are obtained by maximizing the value of using coal or shifting to renewables subject to stochastic prices of coal. From the options values, the trigger price is determined as the minimum price of coal that maximizes the option values between the initial period and the terminal period of investment. At this price of coal, switching to renewables is optimal.

2.3.1   Baseline result

The result of the dynamic optimization is shown in Figure 2.1. The first point of interest is the line at the bottom of each option values curve. These lines represent the net present values of each renewable energy sources as summarized in Table 2.1. The positive net present values indicate positive returns for all types of investment. Among the renewable energy sources, geothermal shows to be the most profitable followed by wind, hydropower, and lastly, solar PV. This is in line with the previous studies showing geothermal to be the cheapest form of energy and most attractive renewable energy investment among all other sources available in the country (Stich and Hamacher (2016); DOE (2014); Utama et al. (2012)). Meanwhile, the data from DOE further show that majority of renewable sources in the country is coming from both geothermal (12% of total generation) and hydropower (9% of total generation). Currently, the country is the worlds second largest producer of geothermal energy second to the USA (Sovacool (2010)). The use of new renewable energy sources including solar and wind energy has also dramatically increased in the recent years because of the favorable renewable energy policy framework and investments from the private sector (IRENA (2017)). However, it should be noted that NPV is not the sole determinant of investments in a real options approach as the option values and optimal timing that maximize the value of investment opportunity must also be accounted for (Dixit and Pindyck (1994)).

The next point of interest is the option values of various RES as shown in Figure 2.1.

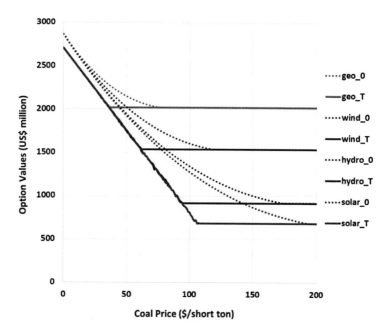

Figure 2.1: Option values of various renewable energy sources
Note: geo_0, geo_T, wind_0, wind_T, hydro_0, hydro_T, solar_0, solar_T are option values for
investment in geothermal, wind, hydropower and solar PV at initial (0) and terminal (T)
period of investment
source: Agaton (2018c)

Table 2.1: NPV of investment in various renewable energy sources

	Net Present Value
Geothermal	US$ 2020M
Wind	US$ 1538M
Hydroelectric	US$ 917M
Solar PV	US$ 681M

Note: The values shown are estimated NPV for substituting 2165GWh/year of electricity generated from coal with various energy sources for 30 years
source: Agaton (2018c)

Each point on the curves describes the maximized value of investment for every price of coal. It can be observed that the option values are greater or equal than the NPV. This is because the real option value equals the net present value of an investment plus the value of management flexibility (Santos et al. (2014); Yeo and Qiu (2003); Trigeorgis (1996)). This highlights the advantage of using ROA over traditional project valuation methods as it combines uncertainty and risk with flexibility while considering the volatility in investment as a potential positive factor which gives additional value to the project (Brach (2003)). Among the option value curves in Figure 2.1, investment in geothermal energy showed the highest real option values followed by wind, hydroelectric, and solar PV. This implies that among various RES options, geothermal seems to be the most attractive investment.

Another point of interest is the dynamics of option values. The downward slope of

the curves indicates that option values decrease with prices of coal. The point where
option value curve meets the straight line is the point of indifference. This point
indicates the price of coal where an investor is indifferent between a decision to opt for
coal or renewables. After such point, an investor has no better option but to switch
technologies to renewables to avoid welfare loses. The vertical distance between the
option curve at the terminal period (broken curve) and initial period (bold curve)
represents the benefit of the option to wait which is equal to ($V_T(P_{C,t}) - V_0(P_{C,t})$).
It can be observed that as coal price increases from zero, the value of option to wait
increases, then decreases, and finally equals zero. This only confirms that the option
value and value of option to wait are not necessarily proportional with changes in coal
price under the conditions set in the real options model.

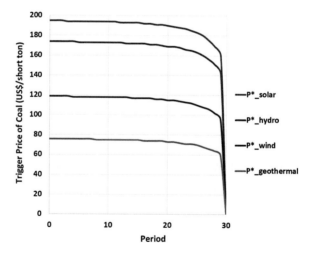

Figure 2.2: Trigger coal price dynamics for various renewable energy sources
Note: P_solar is the trigger price of coal for shifting energy source from coal to solar PV at
each period of investment; P_hydro is the trigger prices for hydroelectric source; P_wind for
wind energy; and P_geothermal for geothermal energy
source: Agaton (2018c)

The last point of interest is the optimal trigger price of coal for shifting technologies and the dynamics of trigger prices over time as shown in Figures 2.1 and 2.2. From Figure 2.1, the optimal trigger price is represented by the intersection of the option values curves between the terminal and the initial period of investment. At this point, the value of waiting to invest becomes zero, which signifies the optimal decision to shift technologies to renewables. The trigger prices for various renewable energy sources are US$74 for geothermal, US$116 for wind, US$173 for hydropower, and US$194/short ton for solar. This only conforms the NPV result in table 1 which implies geothermal as the most attractive renewable energy option in terms of the profit from investment under uncertainty in coal prices. Figure 2.2 also follows this conclusion as geothermal yields the lowest trigger price curves at every period of investment.

2.3.2   Sensitivity in discount rate and volatility of coal price

This portion of analysis describes how optimal strategies adjust with varying certain investment parameters. Table 2.2 shows the trigger prices of coal increases at different discount rates. As discount rate increases, the net present value of investment decreases due to a decrease in net cash flow after discounting. Lower net present values result to a lower option values thereby decreasing the benefit of switching technologies to renewables. Thus, higher discount rate increases the trigger price implying a more optimal decision to wait or to delay investment in renewables. This suggests that the government must set a discount rate lower than the current 10 to 15% to attract investors to renewable energies and add benefit to power producers to shift energy sources from coal to renewables. This result conforms with previous studies affirming that RES with higher proportion of capital investments are more expensive with high discount rate, while electricity generated from coal (and other fossil fuels) with

significant proportion of operational and variable costs are far less susceptible to changes
in discount rates due to lower capital costs (Copiello et al. (2017); Jeon et al. (2015);
Pereira-Jr et al. (2013)). Hence, lower social discount rates entail greater contributions
of renewable technologies while higher social discount rates favor the use of fossil fuels
(Gusano et al. (2016)).

Table 2.2: Optimal trigger prices of coal at various discount rates

	Discount Rate (US\$/short ton)			
	5%	7.5%	Base (10%)	12.5%
Geothermal	\$41	\$58	\$74	\$97
Wind	\$85	\$92	\$116	\$152
Hydroelectric	\$119	\$136	\$173	\$218
Solar PV	\$128	\$152	\$194	\$224

Note: Values shown are optimal trigger prices of coal, in US\$/short ton, for shifting energy
source from coal to various renewable sources at different discount rates: 5%, 7.5%,
10%(baseline value), 12.5%
source: Agaton (2018c)

Table 2.3 shows the trigger prices at different volatilities in coal prices. The results show
that at lower volatility of coal prices, option values increase, and the optimal trigger
price of shifting to renewable decreases. This implies that with more deterministic
prices of coal input, the better to invest earlier in renewables. On the other hand,
when prices of coal input are more volatile in the market, it is better to wait and delay
the investment to avoid losses from investment risks. This is because investors are risk
averse, and they accept a riskier project only if they expect to receive a higher return

to compensate it (Viallet and Hawawini (2015)). Therefore, the higher the uncertainty over future cash flows, the lower the projects NPV and real option value, and lesser attractive to investors (Fagiani et al. (2013)).

Table 2.3: Optimal trigger prices of coal at various coal price uncertainties

	Coal Price Uncertainty (US$/ short ton)		
	Low	Base	High
Geothermal	$61	$74	$85
Wind	$95	$116	$133
Hydroelectric	$142	$173	$198
Solar PV	$159	$194	$222

Note: Values shown are optimal trigger prices of coal, in US$/short ton, for shifting energy source from coal to various renewable sources at different levels of coal price uncertainty(σ): low uncertainty=0.1; baseline=0.225762; high uncertainty=0.3
source: Agaton (2018c)

2.4 Limitations and discussion

To develop a ROA framework of energy investment decision, this study made several simplifying assumptions leading to various limitations in the analyses. First, the coal prices are assumed to be stochastic and follow GBM. With the positive drift of coal prices from the ADF unit root test, this assumes that the prices are increasing in the long run. With coal being an exhaustible resource, the current demand path and competition may accelerate the upward trend in prices with higher volatility and uncertainty. On the other hand, this paper acknowledges that the rapid developments in renewable technologies may eventually reduce the demand for coal and its price. This trend in coal price and demand should be accounted for. Moreover, different models to describe stochastic prices of fossil fuels could also be used for further comparison of results using GBM.

The study applies ROA under uncertainty in fuel prices and social discount factor. This paper acknowledges other uncertainties that affect investment decisions particularly to renewable energy. These include inflation that affects estimates of future cash flow; changes in FiT rates and market price of electricity; FiT subsidies and other government incentives; technological innovation that could lower the investment cost for RES; and environmental policy by imposing externality tax and green energy certificates (Eissa and Tian (2017); Kitzing et al. (2017); Zhang et al. (2017); Byrnes et al. (2016); Eryilmaz and Homans (2016); Kamjoo et al. (2016); Arnold and Yildiz (2015)). The proposed ROA could be extended by incorporating these uncertainties to make a better investment strategy and comparison among various energy sources.

This study focuses only on the financial side of renewable energy investment. In real project valuation and decision-making process, there are also several factors considered including economic impacts on employment, electricity prices, and local economy; environmental impacts on landscape, wildlife, noise level and quality of air; and socio-technical factors including users energy demand, usage patterns, system sizing, and availability of renewable resources (Barros et al. (2017); Akinyele and Rayudu (2016); Akinyele et al. (2015); Emmanouilides and Sgouromalli (2013)). Future studies could incorporate these factors to make the current ROA model more robust and valuable not only to investors but to project evaluators and policy makers as well.

Although there are some limitations, the ROA framework proposed in this study could be a good benchmark for further analysis of investment decisions for cleaner and more sustainable sources of energy.

2.5   Conclusion

This study presented investment environments for switching coal-based electricity generation to renewable energy by incorporating the option to delay or to wait in making an investment decision. Using MATLAB programming, dynamic optimization processes were done to evaluate the option values of investments and the timing switching technologies. Sensitivity analyses were performed to identify how uncertainty in coal prices and discount rates affect the investment decision-making process. By applying the real options approach, this study characterized scenarios where renewable energy became more attractive option than continue using coal. Among the renewable sources, geothermal energy showed to be the most profitable option followed by wind, hydroelectric, and solar PV. More deterministic coal prices and lower discount rates decreased the trigger prices of coal suggesting an earlier shifting of technologies from coal to renewables. Furthermore, higher coal price uncertainty and higher discount rate indicated a more optimal decision to delay investment in renewable energy.

2.6   Appendix

Table 2.4: Summary of Estimated Parameters

Parameter	Value	Unit	Description
alpha	0.35094		estimated myu value of GBM unit root test of coal prices
sigma	0.22576 2		standard deviation of GBM unit root test of coal prices
rho	0.91		discount factor
Pmin	0	US$/short ton	base-level for price of coal
Pmax	200	US$/short ton	maximum limit for price of coal
Pstep	1	US$/short ton	the value between each price node
P_e	182.2	US$/MWh	price of electricity
Q_e	2165150	MWh	average annual electricity generated
Q_i	1324562	short ton	average annual quantity of coal used to generate Q_e
C_c	40.7M	US$	annual O&M cost to generate Q_e from coal
C_g	28.8M	US$	annual O&M cost to generate Q_e from geothermal
C_w	28.4M	US$	annual O&M cost to generate Q_e from wind
C_h	18M	US$	annual O&M cost to generate Q_e from hydro
C_s	21.9M	US$	annual O&M cost to generate Q_e from solar PV
I_g	1.99B	US$	investment cost of geothermal to generate Q_e
I_w	2.31B	US$	investment cost of wind to generate Q_e
I_h	3.61B	US$	investment cost of hydro to generate Q_e
I_s	4.13B	US$	investment cost of solar to generate Q_e
LL		years	time horizon for dynamic optimization problem
T_c		years	number of periods for NPV of coal
T_r		years	time horizon for renewable energy generation
5%, 7.5%, 10%, 12.5%			social discount rates
0.1, 0.225762, 0.3			uncertainty levels (low, base, high)

Note: The table summarizes the estimated parameters used in dynamic optimization to compute the option values for each renewable energy source.
source: Agaton (2018c)

Table 2.5: ADF Unit Root Test result for coal prices(1984-2016)

Null Hypothesis: LNP has a unit root
Exogenous: Constant
Lag Length: 3 (Automatic - based on SIC, maxlag=9)

		t-Statistic	Prob.*
Augmented Dickey-Fuller test statistic		-1.395821	0.5723
Test critical values:	1% level	-3.646342	
	5% level	-2.954021	
	10% level	-2.615817	

*MacKinnon (1996) one-sided p-values.

Augmented Dickey-Fuller Test Equation
Dependent Variable: D(LNP)
Method: Least Squares
Date: 08/23/17   Time: 15:30
Sample (adjusted): 1984 2016
Included observations: 33 after adjustments

Variable	Coefficient	Std. Error	t-Statistic	Prob.
LNP(-1)	-0.109118	0.078175	-1.395821	0.1737
D(LNP(-1))	0.211469	0.168819	1.252635	0.2207
D(LNP(-2))	-0.322127	0.158554	-2.031649	0.0518
D(LNP(-3))	0.537105	0.170162	3.156427	0.0038
C	0.439470	0.297489	1.477266	0.1508

R-squared	0.363552	Mean dependent var	0.035094
Adjusted R-squared	0.272630	S.D. dependent var	0.225762
S.E. of regression	0.192543	Akaike info criterion	-0.318268
Sum squared resid	1.038038	Schwarz criterion	-0.091524
Log likelihood	10.25142	Hannan-Quinn criter.	-0.241975
F-statistic	3.998536	Durbin-Watson stat	1.648843
Prob(F-statistic)	0.010896		

Note: Own estimation using coal prices from 1984-2016.
source: Agaton (2018c)

Chapter 3

Coal, renewable, or nuclear? A real options approach to energy investments in the Philippines

Abstract - This chapter is based on a article published in *International Journal of Sustainable Energy and Environmental Research*[4]. The aim of this paper is to evaluate the comparative attractiveness of either investing in alternative energy sources or continuing the use of coal for electricity generation in the Philippines. Applying the real options approach under coal price uncertainty, this study analyzes investment values and optimal timing of switching technologies from coal to renewable or nuclear energy. It also examines how negative externality and the risk of nuclear accident affect investment decisions. Results identify possible welfare losses from waiting or delaying investing in alternative energy. Negative externality favors investment in nuclear energy over coal, whereas the risk of nuclear accident favors investment in renewable energy.

3.1 Introduction

The rapid economic development in the Philippines causes dramatic increase in country's energy demand in the recent decades. As the country's electricity sector is highly dependent on imported coal as a source of energy for power generation, the country's

---

[4]Agaton (2017a) available at `https://doi.org/10.18488/journal.13.2017.62.50.62`

energy security has been vulnerable to various crises and unstable coal prices. To
address the increasing energy demands and decreasing dependence on imported coal,
the government started its nuclear program during the world oil crisis in 1973. However,
due to numerous protests related to nuclear disasters, controversies, and nuclear safety,
the succeeding administration discontinued the program (Beaver (1994)). In the recent
years, the government is considering to rehabilitate the suspended plant and construct
four additional nuclear power plants as a long-term option for energy source in the
country (International Atomic Energy AgencyIAEA (2016)). Renewable energies,
on the other hand, remain the most promising alternatives to suffice the country's
energy demand. At present, renewable energy sources, particularly geothermal and
hydropower, account to 25% of the country's power capacity (DOE (2016a)). The
country is aiming to increase this capacity to 60% and become energy independent
by 2030 by developing localized renewable energy resources (DOE (2012)). However,
competitive prices of coal, economic downturns, political instability, natural calamities,
and skepticism challenge the investments on these alternatives. This study takes this
motivation to suggests a strategy whether to invest or not, and when to invest on
alternative energy source to address the country's problem on energy security and
sustainability.

Recent studies discuss renewable energy investments in the Philippines. Hong and Abe
(2012) use multiple correspondence analysis to deal with the technical, economic, and
social aspects of developing renewable energy projects to promote energy sustainability;
Meller and Marquardt (2013) present a holistic approach to calculate the costs of RE
and compare their competitiveness with conventional sources of fuel; and Sovacool
(2010) proposes an analytical framework to evaluate RE support mechanisms such as
renewable portfolio standards, green power programs, public research and development

expenditures, systems benefits charges, investment tax credits, production tax credits, tendering, and feed-in tariffs in Southeast Asia including the Philippines. However, the methodologies in these literatures do not capture important characteristics of investment such as irreversibility, uncertainty, and flexibility in timing of investment Baecker (2007). Real options approach (ROA) overcomes these limitations by combining uncertainty and risk with flexibility of investment as potential factors that give additional value to the project Brach (2003).

Myers (1977) referred the term real options approach (ROA) to the application of option pricing theory to valuate non-financial or real assets. It is useful in project appraisal when revenues from investment contain uncertainty in the future cash flow and when there is a possibility to choose the timing of investment (Yang et al. (2008)). Recent studies use ROA to analyze investment decision particularly with renewable energy. These include Zhang et al. (2016) on the application of real options to solar photovoltaic power generation in China; Kitzing et al. (2017) on the analysis of wind energy investments under different support schemes; and Kim et al. (2017) on analyzing uncertainty variables affecting investment in developing countries with a case in Indonesia. Several studies also use this approach to analyze nuclear energy investments including the works of Rothwell (2006) on evaluating new nuclear power plants in the United States of America; Shi and Song (2013) on evaluating how risks and uncertainties affect the development of new power plants in China; Tian et al. (2016) on analyzing the influence of carbon market on nuclear investment in China; and Cardin et al. (2017) on the flexibility analysis for nuclear power plants with uncertainty in electricity demand and public acceptance. This research tries to contribute to these literatures by analyzing energy switching problem from coal to renewable energy or nuclear energy, involving uncertainty in coal prices, negative externality, and the risk of nuclear accident.

This paper presents a framework of energy investment strategy that applies to developing countries which are highly dependent on imported fuel for electricity generation. The main goal is to provide an example of a framework of full-system switch investment decision by applying the case of the Philippines. Although this acknowledge having diverse options for energy investments in the Philippines, this study only focuses on the problem of switching to renewable energy and nuclear energy in line with the country's long-term energy plan (DOE (2012)) and the Philippine nuclear power development program (IAEA (2016)). Specifically, this study aims to evaluate the option values of energy investment and identify the trigger price of shifting technologies from coal to these alternative energies. This further aims to present investment environments where investing in renewable energy is a better alternative than nuclear. These environments include scenarios where externality and risk of nuclear disaster affect the dynamics of option values of trigger price strategy. This finally aims to recommend various government actions to address environmental problem, supply chain, and national security regarding energy.

3.2   Methodology

This study uses ROA to analyze investment decisions whether to continue using coal for electricity generation or shift to alternative energy sources. Matlab programming is used to (a) generate transition probability matrix that describe stochastic prices of coal, (b) Monte Carlo simulation to calculate the expected net present value of using coal and expected net present value of nuclear energy considering the probability of an accident, and (c) dynamic optimization that maximizes the value of investment at each price of coal from initial period to final period of investment. From this optimization, the trigger prices of coal for shifting technologies from coal to renewable or nuclear are then identified. To describe a more realistic situation where investors, policy makers, and the people are skeptical in investing in nuclear energy due to its risks, this study poses a scenario of the possibility of having a nuclear accident. Finally, negative externality of using various types of energy is incorporated in the ROA model to reflect national energy security and environmental concerns such as water and air pollution, greenhouse gas emission, and ecosystem and biodiversity loss.

3.2.1 Dynamic Optimization

This study adopts the work of Detert and Kotani (2013) on making investment decisions under uncertainty using dynamic optimization. In this research, ROA is used to describe a model of an investor that maximizes the value of investment of either investing in alternative energy or continuing the use of coal for electricity generation as shown in Equation 3.1 (see Table 3.1 for an overview of all model parameters and variables).

$$\max_{0 \leq \tau < T+1} \mathbb{E} \left\{ \left[\sum_{t=0}^{\tau} \rho^t \pi_{c,t} + \rho^{T_c} NPV_{c,t} \left(1 - \mathbb{I}_{\{\tau \leq T\}} \right) \mid P_{c,0} \right] + \{NPV_A\} \left(\mathbb{I}_{\{\tau \leq T\}} \right) \right\} \quad (3.1)$$

where

$$\pi_{c,t} = P_E Q_E - P_{c,t} Q_c - C_c - E_c \quad (3.2)$$

$$NPV_{c,t} = \sum_{t=T}^{T_c} PV_{c,t} = \sum_{t=T}^{T_c} \rho^t \pi_{c,t} = \left(\frac{1 - \rho^{T_c+1}}{1 - \rho} \right) [P_E Q_E - P_{c,t} Q_c - C_c - E_c] \quad (3.3)$$

$$NPV_A = \sum_{m=0}^{T_R} \rho^m \pi_R - I_R - E_R = \left(\frac{1 - \rho^{T_R+1}}{1 - \rho} \right) [P_E Q_E - C_R] - I_R - E_R \qquad (3.4)$$

or

$$NPV_A = \sum_{m=0}^{T_N} \rho^m \pi_N - I_N - E_N = \left(\frac{1 - \rho^{T_N+1}}{1 - \rho} \right) [P_E Q_E - C_{NF} - C_N] - I_N - E_N \quad (3.5)$$

Using dynamic programming, the option value of investment for each period as shown in Equation 3.6.

$$V_t(P_{c,t}) = max \left\{ \pi_{c,t} + \rho \mathbb{E}(V_{t+1}(P_{c,t+1}) \mid P_{c,t}), NPV_A \right\} \qquad (3.6)$$

The option value, $V_t(P_{c,t})$, is calculated by maximizing the investment at each price of coal, $P_{c,t}$ from 0 to US\$500/short ton. Dynamic optimization process is set to 40 years to represent a situation where an investor is given a period to make an investment decision. After such period, he has no other option but to continue using coal for electricity generation. The choice is valued for another 40 years to represent the lifetime of power plant using coal.

From the dynamic optimization results in equation 3.6, the dynamics of option values is analyzed and trigger price is identified. The trigger price in this model is described as the optimal timing for switching technologies from coal to alternative energy as shown in equation 3.7. From the given equation, the trigger price is evaluated as the minimum price of coal where the option value at the initial period equals the terminal period of investment.

$$\widehat{P}_c = min \left\{ P_{c,t} \mid V_0(P_{c,t}) = V_T(P_{c,t}) \right\} \qquad (3.7)$$

Table 3.1: Description of parameters and variables

Variable	Description, unit
V_t	Option value of investment at each price of coal at each period t, US$
NPV_A	Net present value of using coal for electricity generation, US$
NPV_A	Net present value of investing in renewable or nuclear, US$
$\pi_{c,t}$	Profit of using coal for electricity generation, US$
π_R	Profit for investing in renewable energy, US$
π_N	Profit for investing in nuclear energy, US$
P_E	Price of electricity in the Philippines, US$/MWh
$P_{c,t}$	Stochastic price of coal, US$/short ton
Q_E	Quantity of electricity demand from coal, MWh
Q_c	Quantity of coal needed to produce Q_E, short ton
C_c	Annual marginal cost for electricity generation using coal, US$
C_R	Annual marginal cost for electricity generation from renewable, US$
C_N	Annual marginal cost for electricity generation from renewable, US$
C_{NF}	Annual marginal fuel cost for electricity generation from renewable, US$
I_R	Investment cost for renewable energy, US$
I_N	Investment cost for nuclear energy, US$
C_D	Decommissioning cost for closing nuclear power plant, US$
C_A	Cost of nuclear accident, US$
E_c	Externality cost of generating electricity from coal, US$
E_R	Externality cost for renewable energy generation, US$
E_N	Externality cost for nuclear energy generation, US$
T_R	Lifetime of electricity generation from renewable energy, years
T_N	Lifetime of electricity generation from nuclear energy, years
T_C	Lifetime of electricity generation using coal, years
T	Total period of investment, years
τ	Period where investor decides to invest in renewable or nuclear, years
ρ	Discount factor
$\mathbb{1}_{\tau \leq T}$	Indicator equal to 1 if switching to renewable or nuclear energy is made, otherwise, equal to 0
J	Number of times for Monte Carlo simulation process

source: Agaton (2017a)

3.2.2   Geometric Brownian Motion and Monte Carlo simulation

In line with previous studies, (Tietjen et al. (2016); Wang and Du (2016); Xian et al. (2015)), this study assumes that the price of coal is stochastic and follows Geometric Brownian motion (GBM) with a drift. Using discretized specification for GBM, the price of coal as shown in equation 3.8

$$P_{c,t+1} = P_{c,t} + \alpha P_{c,t} + \sigma P_{c,t} \varepsilon_t \tag{3.8}$$

where α and σ are the drift and variance rates of time series of prices of coal (Dixit and Pindyck (1994)). This equation illustrates that the current price of coal depends on its previous price, and the drift and variance of coal prices. Applying the work of Insley (2002), the values of and are estimated using ADF unit root test. The annual average prices of coal from 1970 to 2016 from World Bank-Global Economic Monitor are used to run the ADF test. The result in Table 3.2 implies that the null hypothesis that p_t has a unit root cannot be rejected at all significant levels. Therefore, coal prices conform with GBM. The estimated GBM parameters are $\alpha = 0.011133$ and $\sigma = 0.250153$ and are used to approximate stochastic prices of coal for each investment period $t = 0$ to $t = T_c$ at each initial price of coal from $P_c = 0$ to P_c=US$300/short ton at an increment of US$ 1/ short ton.

This study applies Monte Carlo simulation to estimate the expected NPV of using coal for electricity generation in Equation 1. In this process, the computation of NPV from Equation 3.3 is repeated in a sufficiently large number of $J = 10000$ times to approximate the expected NPV at each initial price of coal and take the average as

Table 3.2: Augment Dickey-Fuller unit root test of coal prices

Test statistic and significance levels for critical values		t-Statistic	Prob*
Augmented Dickey-Fuller test statistic		-2.338239	0.1648
Test critical values:	1% level	-3.581152	
	5% level	-2.926622	
	10% level	-2.601424	

Note: Table shows the summary of ADF unit root test for coal prices 1971-2016.
source: Agaton (2017a)

shown in Equation 3.9.

$$\mathbb{E}\left\{NPV_{c,J} \mid P_{c,0}\right\} \approx \frac{1}{J}\sum_{j=1}^{J}NPV_{c,j} \approx \mathbb{E}\left\{NPV_c \mid P_{c,0}\right\} \tag{3.9}$$

3.2.3 Risk of Nuclear Accident

Recent literatures discuss the probability of nuclear accident using classical probabilistic models, simple empirical approach, Poisson distribution, Poisson Exponentially Weighted Moving Average (PEWMA), Least Squares Monte-Carlo(LSM), and infinite mean model (Rangel and Lvque (2012); Hofert and Wthrich (2013); Kaiser (2012); Zhu (2012)). However, these do not fit with the ROA model described in this study where the decision to invest in nuclear energy is evaluated in an annual basis and so the probability of nuclear accident. This study proposes a ROA model considering a risk of having a nuclear accident. This study assumes that an accident may happen only once, at most, in the entire lifetime of nuclear energy generation. The energy generation terminates once the accident occurs, hence, accident cannot be repeated.

Assuming an independent and identically distributed (i.i.d.) random variable x_i *Bernoulli* for $i = \tau, \tau + 1, , \tau + T_N$, as shown in equation 3.10.

$$x_i = \begin{cases} 0 \text{ with prob. } q(\tilde{t}) \text{ if no disaster} \\ 1 \text{ with prob. } 1 - q(\tilde{t}) \text{ otherwise} \end{cases} \quad where \ q'(\tilde{t}) < 0 \qquad (3.10)$$

Stopping time, $\tilde{t}$, describes the period which nuclear accident happens subject to

$$\tilde{t} = min \{T_N : x_\tau + x_{\tau+1} + ... + x_{\tau+T_N} = 1\} \qquad (3.11)$$

The probability mass function of this Bernoulli distribution over possible outcomes of k, is

$$Pr\left(\tilde{t} = \tau + k\right) = \left(1 + q\left(\tilde{t}\right)\right)\left(q\left(\tilde{t}\right)\right)^k \qquad (3.12)$$

for $k = 0, 1, 2, ..., T_N$.

The accident may happen after the switch to nuclear energy with $\tilde{t}$ equal to $\tau, \tau+1, , \tau+ T_N, ..., \infty$. Then the probability of a having a nuclear accident is $Pr\left(\tilde{t} \leq \tau + T_N\right)$ where

$$Pr\left(\tilde{t} = \tau\right) + Pr\left(\tilde{t} = \tau + 1\right) + Pr\left(\tilde{t} = \tau + 2\right) + ... + Pr\left(\tilde{t} = \tau + T_N\right) = 1 \qquad (3.13)$$

Using Equation 3.12, the Equation 3.13 can be expressed as

$$\left[1 - q\left(\tilde{t}\right)\right] + \left[1 - q\left(\tilde{t}\right)\right]\left(q\left(\tilde{t}\right)\right) + \left[1 - q\left(\tilde{t}\right)\right]\left(q\left(\tilde{t}\right)\right)^2 + ... + \left[1 - q\left(\tilde{t}\right)\right]\left(q\left(\tilde{t}\right)\right)^{T_N} + ... = 1$$
$$(3.14)$$

Then the probability of having no accident in the lifetime of nuclear energy generation is described as $Pr\left(\tilde{t} > T_N\right) = 1 - Pr\left(\tilde{t} \leq \tau + T_N\right)$. Therefore, the probability of having no nuclear accident decreases over time. The reason behind this is the assumption that nuclear plant increases the risk of an accident, as it gets older especially during a continued operation beyond the end of its useful years. The expected net present value of nuclear energy investment as follows

$$\mathbb{E}\left\{NPV_N\right\} = \mathbb{E}\left\{NPV_N\left(1 - \mathbb{I}_{\{\tilde{t}\leq T_N\}}\right) + NPV_N\left(\mathbb{I}_{\{\tilde{t}\leq T_N\}}\right)\right\} \tag{3.15}$$

where $\mathbb{I}_{\{\tilde{t}\leq T_N\}}$ is an indicator function equal to 1 if nuclear accident occurs, otherwise 0. Expanding the equation 3.15 with probability function for nuclear accident at each period gives

$$
\begin{aligned}
\mathbb{E}\left\{NPV_N\right\} = {} & Pr(\tilde{t} > T_N)\left[\sum_{t=\tau}^{\tau+T_N}\rho^t\pi_N - \rho^{\tau+T_N}C_D - \rho^\tau I_N\right]\left(1 - \mathbb{I}_{\{\tilde{t}\leq T_N\}}\right) \\
& + (\tilde{t} = \tau)\left[\rho^t\pi_N - \rho^\tau(C_D + C_A) - \rho^\tau I_N\right] \\
& + Pr(\tilde{t} = \tau + 1)\left[\rho^t\pi_N - \rho^{\tau+1}(C_D + C_A) - \rho^\tau I_N\right] \\
& + Pr(\tilde{t} = \tau + 2)\left[\rho^t\pi_N - \rho^{\tau+2}(C_D + C_A) - \rho^\tau I_N\right] + \dots \\
& + Pr(\tilde{t} = \tau + T_N)\left[\rho^t\pi_N - \rho^{\tau+T_N}(C_D + C_A) - \rho^\tau I_N\right]\left(\mathbb{I}_{\{\tilde{t}\leq T_N\}}\right)
\end{aligned}
\tag{3.16}
$$

Using Monte Carlo simulation, binomial numbers are generated to represent the probability of having no accident. The process is repeated several times (10000) and get the average to estimate the expected probability value $\mathbb{E}[Pr(\tilde{t} = \tau + t)]$ for each period of nuclear energy generation. These estimates are used to calculate the expected net present value of nuclear investment, $\mathbb{E}\left\{NPV_N\right\}$ in equation 3.1 to determine the option values of investment using the dynamic optimization process.

3.2.4   Data and Scenarios

To determine a suitable set of parameter values for the baseline scenario, we use data from Department of Energy (DOE (2016a)), Energy Information Administration (EIA (2017a)), and Nuclear Energy Agency (NEA (2015)). From these sources, the domestic electricity price, and the quantity of electricity generated from coal are determined.

Using the 2015 electricity production from coal of 36,686GWh, the quantity of coal
and the operations and management cost needed to generate this amount of electricity,
as well as the investment costs, annual operations and management costs, and fuel
costs for nuclear and renewable energy are estimated. The decommissioning cost is
incorporated in the investment cost of nuclear energy (NEA (2016a)). The country
data for wind energy is used to represent the renewable energy investment. The number
of years of nuclear energy generation is set to 50 years while 30 years for renewable
energy. The social discount rate is 7.5%. In the base scenario, all externality values
are set to zero to provide an initial estimate of comparison in later scenarios. This also
describes the current situation in the country where externalities from various sources
are not valued.

In the scenario of nuclear accident, the probability is set to 0.01% per year with damage
cost comparatively higher than the values reported at the NEA (2016b). The accident
cost is set greater than the values reported in literature to describe a more realistic
maximum potential for nuclear damages. The last scenario incorporates the externality
cost of electricity generation from various sources. The values used here are in line with
the external cost of generating electricity in the Philippines (Meller and Marquardt
(2013)) and average external costs for electricity generation technologies (European
Environmental Agency EEA (2010)). The externality values are first set to US$6/MWh
for renewable energy, US$1/MWh for nuclear energy to US$1/MWh while zero for
electricity generation from coal (as described in base scenario). The value of externality
for using coal are then adjusted from 0 to US$100/MWh at US$25/MWh increment.
In this scenario describes how increasing externality cost from coal affects the options
values and trigger prices of shifting technology from coal to alternative energies. This
scenario also finds the threshold of externality cost for shifting technologies from coal

to alternatives.

3.3   Results

3.3.1   Baseline scenario

Figure 3.1 shows the result of estimation for the baseline scenario. The curves in Figure 3.1 illustrate the maximized option values of either continuing the use of coal for electricity generation or investing in renewable energy (blue curves) or nuclear energy (red curves). The first point of interest in this figure is that the option values decrease with coal price. This implies that the value of any investment decreases with higher cost of input fuel. Second, the straight line at the end of the curves indicate a situation where there is no better option but to shift technology from coal to renewable or nuclear. These lines also describe the net present values of renewable and nuclear energy investment. The results show that investment in renewable energy gains higher NPV equal to US\$31.525 billion than in nuclear energy with US\$30.880 billion. However, note that NPV is not the sole determinant of investments in a real options approach. We must also account for the optimal timing that maximize the value of investment opportunity (Dixit and Pindyck (1994)).

The optimal timing of investment in this study is described as the trigger price of shifting technologies from coal to renewable or nuclear. In Figure 3.1, the intersection of the two curves, option value at the initial period of investment (V_0) and at the terminal period (V_T), illustrate the trigger price of coal. At this price, an investor has no better option but to invest in any of the alternative energies. The result in Figure 3.1 shows that the trigger price of coal for investing in renewable is US\$284/short ton

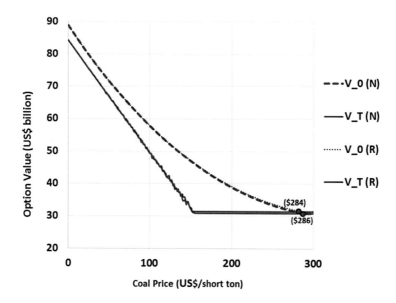

Figure 3.1: Option values of renewable and nuclear energy investments at baseline scenario
Note: V_0(N) -option value of investment in nuclear at initial period, V_T(N) -option value of investment in nuclear at terminal period, V_0(R) -option value of investment in renewable at initial period, V_T(R)-option value of investment in renewable at terminal period
source: Agaton (2017a)

and US$286/short ton in nuclear energy. Although renewable is slightly higher, the difference in trigger prices is not significant. Further, the value of option to wait is described as the difference between the option values at the initial period (dashed curve) and the terminal period. It can observe that option values in the initial period of investment is higher than in the terminal period resulting to a negative value of option to wait in prices of coal below the trigger price. This indicates that waiting to invest in renewable or nuclear energy incurs losses.

3.3.2   Risk of Nuclear Accident Scenario

Figure 3.2 shows the comparison of the option values for nuclear energy investment with the probability of nuclear accident (black curves) and the baseline scenario for both nuclear and renewable. The results reveal that option values of nuclear decrease with the risk of nuclear accident. This result is expected as nuclear accident incurs huge costs to cover the reparation of damages, evacuation of affected residents, rehabilitation, and decommissioning. While this is the case, the trigger price increase from US$286/short ton to US$307/short ton of coal. This marginal increment suggest that it is more optimal to wait longer until the nuclear risk is resolved, or when the nuclear energy technology has advanced to the point of significantly reducing the probability of nuclear accident. Also from the figure, the difference in trigger prices between renewable energy investment and nuclear energy with probability of nuclear accident becomes larger. Further, the options values for renewable is comparably higher than nuclear energy with the risk of accident. This suggests that it more optimal to invest in renewable energy considering the possibility of having nuclear accident.

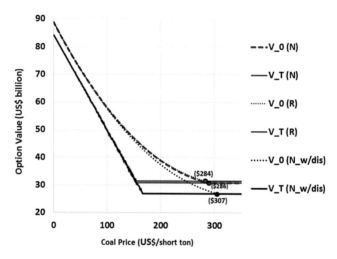

Figure 3.2: Option values with the risk of nuclear accident.
Note: V_0(N) -option value of investment in nuclear at initial period, V_T(N) -option value
of investment in nuclear at terminal period, V_0(R) -option value of investment in renewable
at initial period, V_T(R) -option value of investment in renewable at terminal period,
V_0(N_w/dis) -option value of investment in nuclear at initial period with accident risk,
V_T(N_w/dis) -option value of investment in nuclear at terminal period with accident risk
source: Agaton (2017a)

3.3.3   Negative Externality Scenario

In this scenario analyzes the sensitivity of options values and trigger prices with the addition of negative externality to the base model. Figures 3.3 and 3.4 illustrate the dynamics of options values with negative externality for renewable energy and nuclear energy investments. The results reveal that option values decrease with increasing externality values for both renewable and nuclear energy investments. These results are foreseeable as negative externality incurs additional costs. It is also observed much decrease in the option values for renewable than nuclear energy.

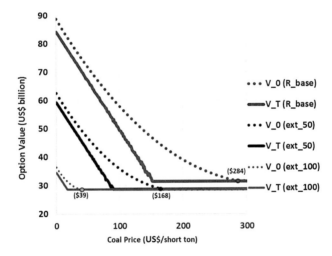

Figure 3.3: Option values of renewable energy investment with negative externality. Note: V_0(R_base) -baseline option value of investment in renewable at initial period, V_T(R_base) -baseline option value of investment in renewable at terminal period, V_0(ext_50) - option value of investment in renewable at initial period with US$50/MWh coal externality, V_T(ext_50) - option value of investment in renewable at terminal period with US$50/MWh coal externality, V_T(ext_100) - option value of investment in renewable at terminal period with US$100/MWh coal externality
source: Agaton (2017a)

Figure 3.5 shows the curves of trigger price of coal for shifting technologies from coal

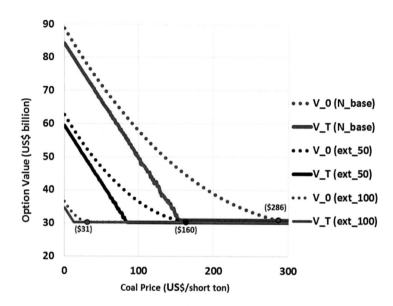

Figure 3.4: Option values of nuclear energy investment with negative externality.
Note: V_0(N_base) -baseline option value of investment in nuclear at initial period,
V_T(N_base) -baseline option value of investment in nuclear at terminal period, V_0(ext_50) -
option value of investment in nuclear at initial period with US$50/MWh coal externality,
V_T(ext_50) - option value of investment in nuclear at terminal period with US$50/MWh
coal externality, V_T(ext_100) - option value of investment in nuclear at terminal period with
US$100/MWh coal externality
source: Agaton (2017a)

to renewable (blue) or nuclear (red) at various externality values of using coal. The results reveal declining trends in the trigger prices for both energy investments with increasing externality values. This suggests that it is more optimal to shift technology earlier from coal to renewable or nuclear energy considering negative externality costs. Also from the figure are the thresholds of externality costs of coal for each investment. This suggests that if the government is eager to attract investors and power producers to shift technologies, the government must set external cost for using coal in a form of externality tax equal to US\$88.93/MWh for renewable energy and US\$85.05/MWh nuclear energy. This result is also in line with the estimated average EU external costs for electricity generation technologies (EEA (2010)).

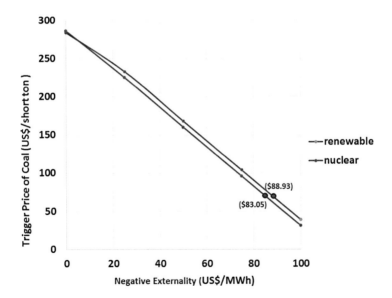

Figure 3.5: Trigger prices of renewable and nuclear energy investment at various negative externalities of using coal
source: Agaton (2017a)

3.4   Conclusion

This study examined various scenarios that represent energy switching investment decisions that apply to developing countries. Although numerous studies explore the effect of input price uncertainty in investment decisions, this study expands the existing body of research by considering switching options to nuclear or renewable energy, incorporating negative externality for using various types of energy, and the risk of nuclear accident.

This study used real options approach to evaluate the option values that maximize the net present value of each alternative investment, and the trigger prices of coal for shifting technologies from coal to renewable or nuclear. Dynamic optimization results showed that flexibility in decision timing is important in making irreversible investment under uncertainty. This highlights the important characteristic of real options approach in valuing financial options that timing is essential in considering investment decisions. Despite the risk of having nuclear accident, investment in nuclear energy seemed to be attractive in the Philippines. Yet, the question on building new nuclear power plant will still be highly debatable as Filipino people are still skeptical from its radiation and health risks due to the recent nuclear accident in Fukushima in 2011. With the long-term reliability, nuclear energy may only serve as a transition technology from coal to renewable as the concerns of the public about safety issues, proliferation of nuclear material, long-term nuclear waste disposal, and risks of using nuclear energy needs to be considered first. Finally, the inclusion of externality cost for using coal makes the option for renewable or nuclear energy investments more valuable than continue using coal for electricity generation. Being nonrenewable and exhaustible, the concerns on coals limited supply, price volatility, national security problems, and the environmental effects associated with its continued use serve as an impetus of finding better and more sustainable sources of energy.

To develop a general model of energy investment decision in developing countries, the study made several simplifying assumptions leading to various limitations in the analyses. It is therefore important to note that the given estimates must be taken with great caution. While the assumptions in this study are sufficient for the main objective of providing qualitative guidance and general scenario of energy investment, it should be noted that thorough identification of parameter estimations requiring calculations with more tailored numerical methods are necessary in real decision-making process. This research focus on real options approach under uncertainty in coal prices, negative externality, and risk of nuclear accident. Future studies may consider other uncertainties associated with energy investments. These include technological innovation that may lower the overnight cost for renewables and safer nuclear energy generation; environmental uncertainty such as climate variability and weather disturbances that affect energy systems; and policy uncertainty to further, capture the underlying political and environmental processes essential to climate change policy.

Chapter 4

A real options approach to renewable electricity generation in the Philippines

Abstract - This chapter is based on an article published in *Energy, Sustainability and Society*[5] journal . This chapter evaluates the attractiveness of investing in renewable energy sources over continue using oil for electricity generation. This paper uses the real options approach to analyze how the timing of investment in renewable energy depends on volatility of diesel price, electricity price, and externality for using oil. The result presents a positive net present value for renewable energy investment. Under uncertainty in oil prices, dynamic optimization describes how waiting or delaying investment in renewables incur loses. Decreasing the local electricity price and incorporating negative externality favor investment in renewable energy over continuing the use of oil for electricity generation. Real options approach highlights the flexibility in the timing of making investment decisions. At the current energy regime in the Philippines, substituting renewable energy is a better option than continue importing oil for electricity generation. Policies should aim at supporting investment in more sustainable sources of energy by imposing externality for using oil or decreasing the price of electricity.

---

[5]Agaton and Karl (2018) available at `https://doi.org/10.1186/s13705-017-0143-y` licensed under *Creative Commons Attribution License 4.0*

4.1   Background

Environmental problems associated with emissions from fossil fuel, along with limited supply, volatile prices, and energy security, prompted developed and developing countries to find more reliable and sustainable sources of energy. Renewable energy (RE) sources, being abundant, inexhaustible, cleaner, and readily available, emerge as a promising alternative energy source. According to International Energy Agency (IEA), RE accounted to 13.7% of the world energy generation mix in 2015 (IEA (2017)). With a rapid decline in RE costs, this percentage mix is expected to double by 2040 (BNEF (2017b)). In the Philippines, the development and optimal use of RE resources is an essential part of the country's low emission strategy and is vital to addressing climate change, energy security, and access to energy (DOE (2012)). In 2015, renewable energy accounts to 25% of the country's total electricity generation mix, only 1% from wind and solar energy (DOE (2016a)). Since the country is highly dependent on imported fossil fuels, sudden changes in the price of fuels in the world market may eventually affect the country's energy security. Renewable energy serves as a long-term solution by introducing localized RE sources. However, despite the country's huge potential for RE generation, investments in RE projects are challenged by competitive prices of fossil fuels, more mature technology for fossil fuels, and very high investment cost for renewable energy. These give us the motivation to make a study that analyzes the attractiveness of RE investments to address the country's concern on energy sufficiency and sustainability.

One of the most common techniques in analyzing investment projects is the net present value (NPV). This technique is widely used by developers, financial institutions, and government agencies under the condition of definite cash flow. Since RE investment in emerging economies involves high risk from volatile energy prices and changing RE technologies, NPV undervalues investment opportunities and thus considered inappropriate for assessing RE projects in developing countries including the Philippines (Kim et al. (2017)). Real options approach (ROA) overcomes this limitation as it combines risks and uncertainties with flexibility in the timing of investment as a potential factor that gives additional value to the project (Brach (2003)). Recent studies use ROA renewable energy investment particularly for wind, solar photovoltaic (PV), hydropower, concentrated solar power (CSP), and combination (hybrid) of RE with uncertainties in non-RE cost, certified emission reduction (CER), feed-in tariff (FIT), energy production, operations and maintenance (O&M) cost, research and development (R&D) grants, production tax credit (PTC), RE credit (REC), among others (see Table 4.1).

This paper contributes to the existing literature by proposing a ROA framework for analyzing RE projects for developing countries, particularly, island countries that are highly dependent on imported oil for power generation. This study takes the case of Palawan, the largest island-province in the Philippines composed of 1780 islands and islets, that is currently not connected to the national grid and only depend on imported diesel and bunker fuel. Applying ROA, this study aims to evaluate whether investing in RE is a better option than continue using diesel for electricity generation by considering various uncertainties in diesel fuel price, local electricity prices, and imposing externality tax for using diesel. This finally aims to recommend various government actions to address environmental problem, supply chain, and national security regarding energy.

Table 4.1: Summary of ROA in literature

Author (Year)	Country	Type	Uncertainty
Detert and Kotani (2013)	Mongolia	hybrid	non-RE cost
Lee et al. (2013)	Indonesia	hydro	CER price
Abadie and Chamorro (2014)	United Kingdom	wind	FiT, energy production, subsidy
Kim et al. (2014)	Korea	wind	non-RE cost
Jeon et al. (2015)	Korea	hydro	FiT, energy production, interest rate, risk free rate, exchange rate
Weibel and Madlener (2015)	Germany	hybrid	energy production, FiT, investment cost
Wesseh-Jr. and Lin (2015)	Liberia	hybrid	non-RE price, R&D funding
Barrera et al. (2016)	Europe	CSP	R&D grant
Eryilmaz and Homans (2016)	United States	wind	PTC, REC
Ritzenhofen and Spinler (2016)	Germany	wind	FiT
Zhang et al. (2016)	China	PV	non-RE cost, FiT, investment cost
Kim et al. (2017)	Indonesia	hydro	energy production, FiT, CER, O&M cost
Kitzing et al. (2017)	Europe	wind	energy price, wind speed
Tian et al. (2017)	China	PV	investment cost

source: Agaton and Karl (2018)

4.2 Methods

4.2.1 Real options approach

Myers (1977) first referred ROA or real options valuation as the application of option pricing theory to valuate non-financial or real assets. Real option itself is as the right, but not the obligation, to take an action (e.g., deferring, expanding, contracting or abandoning) at a predetermined cost, called exercise price, for a predetermined period of time the life of the option (Copeland and Antikarov (2003)). Investment decisions, in the real world has main characteristics: irreversible, high risk and uncertain, and flexible (Baecker (2007)). These characteristics are not captured by traditional methods of valuation, such as NPV, discounted cash flow (DCF), internal rate of return (IRR), and return on investment (ROI) leading to poor policy and investment decisions. ROA, on the other hand, combines uncertainty and option flexibility which characterize many investment decisions in the energy sector.

This research applies ROA to analyze investment decisions whether to continue using diesel for electricity generation or invest in RE. We use the uncertainty in diesel prices as a main factor that affects investment decisions. Using dynamic optimization, we evaluate the maximized value of investment at each price of diesel, identify the trigger price for shifting technology from diesel-based electricity to RE, and analyze the value of waiting or delaying to invest in RE. Finally, we incorporate sensitivity analyses with respect to electricity prices and externality tax for using diesel.

4.2.2 Dynamic Optimization

We follow the method described by Dixit and Pindyck (1994) and adopt the work of Detert and Kotani (2013) on optimizing investment decision under uncertainty using dynamic programming. In this research, we describe a model of an investor that identifies the optimal value of either investing in renewable energy or continue using diesel for electricity generation as shown in eq. (4.1).

$$
V_{D,t} = \max_{0 \leq \tau < T+1} \left[\left\{ \sum_{t=0 \leq t < \tau} \rho^t \pi_{D,t} + \rho^{T_D} \mathbb{E} NPV_{D,t} \left(1 - \mathbb{I}_{\{\tau \leq T\}} \right) \right\} \mid P_{c,0} + NPV_R \left(\mathbb{I}_{\{\tau \leq T\}} \right) \right]
$$
$$(4.1)$$

where

$$
\pi_{D,t} = P_E Q_E - P_{D,t} Q_D - C_D - E_D \tag{4.2}
$$

$$NPV_{D,t} = \sum_{t=T}^{T_D} PV_{D,t} = \sum_{t=T}^{T_D} \rho^t \pi_{D,t} \qquad (4.3)$$

$$NPV_R = \sum_{t=\tau}^{T_R} PV_{R,t} = \sum_{t=\tau}^{T_R} \rho^t \pi_{R,t} = \left(\frac{1 - \rho^{T_R+1}}{1 - \rho} \right) [P_E Q_E - C_R] - I_R \qquad (4.4)$$

Using this model, we determine the option value, $V_{D,t}$, by maximizing the investment at each price of diesel, D, from 0 to US\$1000/barrel, for each investment period, t. We set the dynamic optimization process to 40 years which represent a situation where an investor is given a period to make an investment decision. After that period, he has no other option but to continue using diesel for electricity generation. The choice is valued for another 25 years to represent the lifetime of power plant using diesel. We set the value of T_R to 25 years to represent the number of years of electricity generation using renewable energy. Finally, we solve the problem backwards using dynamic programming from terminal period (Bertsekas (2012); Detert and Kotani (2013)). The uncertainty in diesel prices in equations 4.2 and 4.3 as well as the Monte Carlo simulation in the dynamic optimization process are discussed in the next subsection.

4.2.3   Stochastic Prices and Monte Carlo simulation

In line with previous studies, we assume that the price of diesel is stochastic and follow Geometric Brownian Motion (GBM) (Postali and Picchetti (2006); Guedes and Santos (2016); Fonseca et al. (2017)). Dixit and Pindyck (1994) presents the stochastic price process as

$$\frac{dP}{P} = \alpha dt + \sigma dz \qquad (4.5)$$

where α and σ represents the mean and volatility of diesel price, dt is the time increment, and dz is the increment of Wiener process equal to $_t\sqrt{dt}$ such that $\varepsilon_t \sim N(0,1)$. Using

Table 4.2: ADF unit root test of oil prices

		t-Statistic	Prob
Augmented Dickey-Fuller test statistic		-1.5109	0.5168
Test critical values:	1% level	-3.6268	
	5% level	-2.9458	
	10% level	-2.6115	

Note: Own calculation based on oil prices from 1981-2016. Full ADF unit root test result in Appendix Table 4.5.
source: Agaton and Karl (2018)

Itos lemma, we arrive at $F(P) = lnP$ and

$$dF = \alpha dt + \sigma dz - \frac{1}{2}\sigma^2 dt \qquad (4.6)$$

We approximate equation 4.6 in discrete time as

$$p_t - p_{t-1} = \left(-\frac{1}{2}\sigma^2 \right) t + {}_t\sqrt{dt} \qquad (4.7)$$

To determine the drift and variance of P, we use Augmented Dickey-Fuller (ADF) unit root test using the following regression equation

$$p_t - p_{t+1} = c(1) + c(2)p_{t-1} + \sum \lambda_j y_{t-j} + e_t \qquad (4.8)$$

where $c(1) = (\alpha - \frac{1}{2}\sigma^2)\Delta t$, and $e_t = \sigma\varepsilon_t\sqrt{\Delta t}$. We then estimate the maximum-likelihood of the drift $\alpha = \mu + \frac{1}{2}s^2$ and variance $\alpha = s$, where α is the mean and s is the standard deviation of the series $p_t - p_{t+1}$ (Insley (2002)).

In this research, we use the annual prices of diesel from 1980 to 2016. The result of ADF test as shown in table 4.2 implies that the null hypothesis that p_t has a unit root at all significant levels cannot be rejected. Therefore, P conforms GBM. We estimate the parameters $\alpha = 0.007614$ and $\sigma = 0.358889$ and use in identifying stochastic prices of diesel under GBM.

We use the Monte Carlo simulation to compute the expected net present value of electricity generation using diesel in Equations 4.2 and 4.3. First, we approximate a vector of potential prices of diesel using the stochastic prices of GBM as follows:

$$P_{D,t} = P_{D,t-1} + \alpha P_{D,t-1} + \sigma P_{D,t-1} \epsilon_{t-1} \tag{4.9}$$

This equation illustrates that the previous price affects the current price of diesel. Second, from the initial price of diesel, $P_{D,0}$, we estimate the succeeding prices of diesel in each period using equation 4.9. We incorporate these prices in the equation 4.2 and calculate the present values of using diesel for electricity generation. Finally, we estimate the expected net present value at each initial price node i and repeat the whole process in a sufficiently large number of $J = 10000$ times and take the average as given by the equation

$$\mathbb{E}\left\{NPV_{D,J} \mid P_{D,0}\right\} \approx \frac{1}{J}\sum_{j=1}^{J} NPV_{D,j} \approx \mathbb{E}\left\{NPV_D \mid P_{D,0}\right\} \tag{4.10}$$

4.2.4   Trigger Price Strategy

Dynamic optimization process in the previous sections generates the maximized option values of investment. From these simulation results, we identify the trigger price of diesel for switching to renewable energy as follows

$$\widehat{P_D} = min\left\{P_{D,t} \mid V_0(P_{D,t}) = V_{T_R}(P_{D,t})\right\} \tag{4.11}$$

where $\widehat{P_D}$ is the trigger price of diesel or the minimum price where the option value in the initial period $V_0(P_{D,t})$ is equal to the option value in the terminal period of investment $V_{T_R}(P_{D,t})$ (Dixit and Pindyck (1994); Davis and Cairns (2012); Detert and Kotani (2013)). From the given equation, we define trigger price as the minimum price of diesel that maximizes the profit of shifting the source of electricity from diesel power plant to renewable energy.

Table 4.3: Description of parameters and variables

Notation	Description
$V_{D,t}$	Option value of investment at each price of diesel, D, at each period of investment, t, in US$
$\mathbb{E}NPV_{D,t}$	Expected net present value of continuing diesel for electricity generation, in US$
NPV_R	Net present value of investing in renewable energy, in US$
$\pi_{D,t}$	Profit of using diesel for electricity generation from initial period of investment, 0, to period of switching to renewable energy, τ, in US$
T	Total period of investment
τ	Period of switching from diesel to renewable energy
$\mathbb{1}_{\tau \leq T}$	Indicator equal to 1 if switching to renewable energy is made, otherwise, equal to 0
ρ	Discount factor
P_E	Electricity price, in US$/MWh
$P_{D,t}$	Stochastic price of diesel, in US$/barrel
Q_E	Quantity of electricity produced, in MWh
Q_D	Quantity of diesel needed to produce Q_E, in barrels
C_D	Annual marginal cost of electricity production using diesel, in US$
C_R	Annual marginal cost of electricity production using renewable energy, in US$
I_R	Investment cost for renewable energy, in US$

source: Agaton and Karl (2018)

4.2.5   Data and Scenarios

To determine a suitable set of parameter values for the baseline scenario, we use data from various sources that nearly reflects the investment environment for renewable energy project in Palawan. We set the recent Calatagan Solar Farm project in Batangas as a benchmark of the data for investment in renewable energy (DOE (2016a)). This 63.3MW solar farm, covering a total area of 160 hectares, projects to generate 88620MWh of electricity per year. It costs US\$ 120 million and will operate for at least 25 years. We use the data from Palawan Electric Cooperative (Paleco (2016)) to approximate the local electricity price, and the quantity and costs of generating electricity from diesel.

Electricity prices in the Philippines varies from island to island depending on the source of energy, as well as various charges including the generation, transmission, distribution, metering, loss, among others. In Palawan, effective power rates also varies across different municipalities (Paleco (2016)). We employ these variations in the electricity price scenario by changing the electricity price in the baseline model. In this scenario, we aim to describe how policy in imposing electricity price ceiling or price floor affects the investment decisions particularly in introducing renewable energy as a source for electricity generation.

Lastly, we consider the externality cost of electricity generation from diesel. This includes, but not limited to, health and environmental problems associated with combustion of diesel. We use the data of the estimated average external costs for electricity generation technologies from European Environmental Agency (EEA (2010)). For this scenario, we include externality costs for estimating the net present value of using diesel in Equations 4.2 and 4.3. We arbitrarily assign values, between 0 (for baseline) to US$ 80/ MWh, which are lower than those reported in literature to describe a more realistic condition. We assume that renewable energy source, particularly solar PV, produces minimal or nearly no externality.

4.3 Results

4.3.1 Baseline scenario

Figure 4.1 and table 4.3 show the result of dynamic optimization at the baseline scenario. The first point of interest is the positive net present value of renewable energy. This implies that, using the traditional valuation method, renewable project is a good investment in the island of Palawan. This result is evident as the installation of solar energy projects grows rapidly in the recent years. In 2016, there are already 538.45MW installed capacity of solar projects from the 4399.71 potential capacity in the whole country (DOE (2016b)). Caution must be applied as net present value is not the sole determinant of investment in a real options approach. The optimal timing that maximizes the value of investment opportunity under uncertainty must also be accounted for (Dixit and Pindyck (1994)).

Figure 4.1 shows the dynamics of the option values at different initial prices of diesel. Result shows that the option values decrease over diesel price as the cost of generating electricity increases with fuel price. The trigger price as indicated by the intersection of option value curves indicates the minimum price of diesel that maximizes the decision of shifting from diesel-based to renewable electricity generation. The result in the baseline scenario at US$168/barrel is higher than the current price at US$101.6/barrel. Initially, this implies that waiting to invest in renewable energy is a better option than

Table 4.4: Summary of Dynamic Optimization at Baseline Scenario

Net Present Value of Renewable Energy	**US\$ 104.97 million**
Trigger price of Diesel	**US\$ 168 million/barrel**
Option Value at initial period (at current diesel price)	**US\$ 141.38 million**
Option Value at terminal period (at current diesel price)	**US\$ 104.97 million**
Value of Waiting (at current diesel price)	**US\$ 36.41 million**

Note: The values shown are estimates from substituting 88.6GWh/year of electricity generated from bunker fuel with solar PV

source: Agaton and Karl (2018)

investing at the current price of diesel. However, the value of waiting to invest as describe by the distance between options value curves from initial to terminal period is negative. As seen in Table 4.4, the option value at the current price of diesel at the initial period of investment is US\$ 141.38 million and decreases to US\$104.97 million at the terminal period. This results to a US\$36.41 million loss from delaying or waiting to invest. This implies that waiting to invest in renewable energy incurs losses.

4.3.2   Electricity Price Scenario

This scenario describes how adjusting the local electricity price affects the option values and the trigger price. Figures 4.2 and 4.3 show the dynamics of option values with increasing and decreasing electricity prices. Result shows that the option values shift upwards with increasing electricity prices. This shows that at higher electricity prices, the value of either renewable energy or diesel-based electricity both increase. However, the trigger prices of diesel also increase to US\$172/barrel at US\$220/MWh

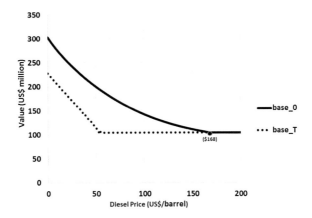

Figure 4.1: Option values at the baseline scenario
Note: base_0 -option values of energy investment at the initial period; base_T -option values
of energy investment at the terminal period
source: Agaton and Karl (2018)

and US$185/barrel at US$250/MWh from the baseline electricity price of US$202/MWh.
This suggests that increasing the electricity price encourage waiting or delaying to
invest in renewable energy.

On the other hand, decreasing electricity prices shifts the option value curves downwards
and decreasing the trigger price of diesel. This result is apparent as decreasing electricity
price results to a lower revenue and thus lower profit for both options. The trigger prices
of diesel decrease to US$160/barrel at US$180/MWh, US$150/barrel at US$150/MWh,
and US$139/barrel at US$120/MWh price of electricity (Figure 4.4). This suggests
that lowering the electricity price decreasing the timing to invest in renewable energy.
Further, the option values become negative at electricity price below US$120/MWh.
This implies that policy makers or power producers must not set an electricity price
below US$120/MWh, as this will result to a loss for producing electricity from diesel
as well as a negative investment for renewable energy.

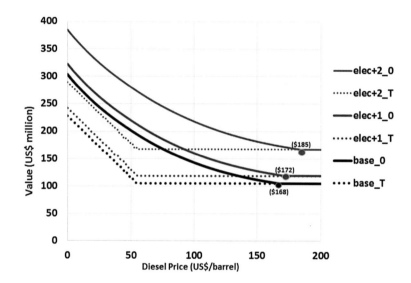

Figure 4.2: Option values at increasing electricity price scenario
Note: base_0 -option values of energy investment at the initial period; base_T -option values
of energy investment at the terminal period; elec+1_0 -option values at 10% higher electricity
price than the base at the initial period; elec+1_T -option values at 10% higher electricity
price than the base at the terminal period; elec+2_0 -option values at 25% higher electricity
price than the base at the initial period; elec+2_T -option values at 25% higher electricity
price than the base at the terminal period
source: Agaton and Karl (2018)

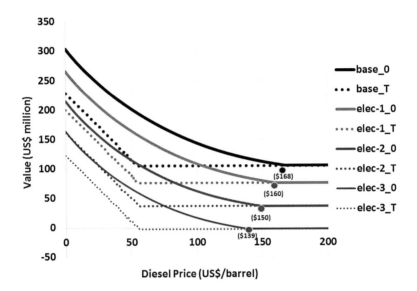

Figure 4.3: Option values at decreasing electricity price scenario
Note: base_0 -option values of energy investment at the initial period; base_T -option values of energy investment at the terminal period; elec-1_0 -option values at 10% lower electricity price than the base at the initial period; elec-1_T -option values at 10% lower electricity price than the base at the terminal period; elec-2_0 -option values at 25% lower electricity price than the base at the initial period; elec-2_T -option values at 25% lower electricity price than the base at the terminal period; elec-3_0 -option values at 40% lower electricity price than the base at the initial period; elec-3_T -option values at 40% lower electricity price than the base at the terminal period
source: Agaton and Karl (2018)

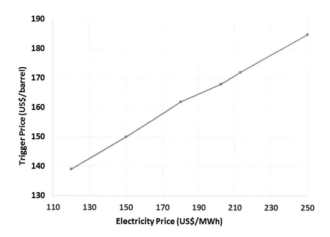

Figure 4.4: Trigger prices of diesel over electricity price
Note: Figure shows the trigger prices of diesel for shifting electricity source from bunker fuel
to solar PV at different domestic prices of electricity.
source: Agaton and Karl (2018)

4.3.3   Negative Externality Scenario

This scenario describes how inclusion of externality tax from combustion of diesel
affects the option values and trigger prices in investment in RE projects. The result in
Figure 4.5 shows that options values shift to the left. First, this implies that imposing
externality tax decreases the revenue from electricity generation using diesel and thus
decreasing the option values. Second, the unchanged lower boundary of the curves
implies externality does not affect the value of investment in renewable energy. This is
due to our assumption that electricity generation from RE produces no externality.

With externality, the trigger prices of diesel decrease to US$140/barrel at US$20/MWh,
US$111/barrel at US$40/MWh, US$82/barrel at US$60/MWh, and US$54/barrel
at US$80/MWh externality cost (Figures 4.5 and 4.6). This implies that imposing

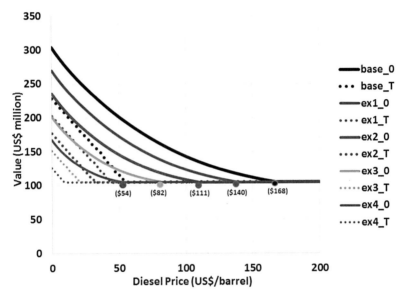

Figure 4.5: Option values at negative externality scenario

Note: base_0 -option values of energy investment with no externality at the initial period; base_T -option values of energy investment with no externality at the terminal period; ex1_0 -option values at 20\$/MWh externality cost at the initial period; ex1_T -option values at 20\$/MWh externality cost at the terminal period; ex2_0 -option values at 40\$/MWh externality cost at the initial period; ex2_T -option values at 40\$/MWh externality cost at the terminal period; ex3_0 -option values at 60\$/MWh externality cost at the initial period; ex3_T -option values at 60\$/MWh externality cost at the terminal period; ex3_0 -option values at 80\$/MWh externality cost at the initial period; ex4_T -option values at 80\$/MWh externality cost at the terminal period;

source: Agaton and Karl (2018)

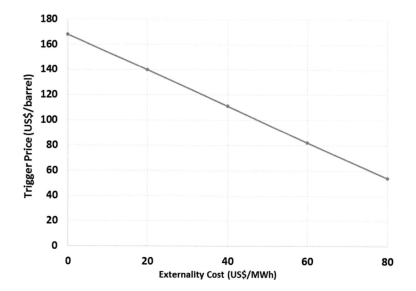

Figure 4.6: Trigger prices of diesel over negative externality
Note: Figure shows the trigger prices of diesel for shifting electricity source from bunker fuel
to solar PV at different externality costs.
source: Agaton and Karl (2018)

externality tax for diesel makes investment in RE more optimal than continue using diesel. Finally, the threshold of externality cost is US$46.55/MWh at the current diesel price of US$101.64/barrel. This is the minimum externality cost that favors immediate investment in RE than continue using diesel.

4.4 Conclusion

We evaluate investment environments and decision-making process for substituting diesel power plant with RE for electricity generation in the Philippines. Using real options approach under uncertainty in diesel prices, we identify the option values, trigger prices of diesel, and value of waiting to invest in RE. We analyze the sensitivity of investment decisions with respect to various electricity prices and addition of externality tax for using diesel. ROA highlights the flexibility in the timing of making investment decisions. Our analyses conclude that, for a developing country that is highly dependent on imported fuel, shifting to RE is a better option than continue using imported diesel. Policies should aim at supporting investment in more sustainable sources of energy by imposing externality for using fossil-based fuel or decreasing the price of electricity. This may negatively affect the power producers but encourage them to shift from diesel to renewable energy.

We summarized a unique approach to energy investment by replacing diesel with RE for electricity generation. We believe that the ROA framework introduced in this research is a good benchmark for further application. First, ROA may take account of environmental and social costs. This may include the cost of deforestation for solar farm, wildlife and habitat loss, air and water pollution, damage to public health, and loss of jobs. Finally, analyzing investment decisions with several RE resources. This includes dynamic optimization with different scenarios of generation mix from various RE sources. We are optimistic that this research becomes one-step forward for further analysis of investment in more sustainable sources of energy.

4.5   Appendix

Table 4.5: ADF Unit Root Test result for Oil Prices (1981-2016)

Null Hypothesis: LNPRICE has a unit root
Exogenous: Constant
Lag Length: 0 (Automatic - based on SIC, maxlag=9)

		t-Statistic	Prob.*
Augmented Dickey-Fuller test statistic		-1.510866	0.5168
Test critical values:	1% level	-3.626784	
	5% level	-2.945842	
	10% level	-2.611531	

*MacKinnon (1996) one-sided p-values.

Augmented Dickey-Fuller Test Equation
Dependent Variable: D(LNPRICE)
Method: Least Squares
Date: 03/16/17   Time: 15:32
Sample (adjusted): 1981 2016
Included observations: 36 after adjustments

Variable	Coefficient	Std. Error	t-Statistic	Prob.
LNPRICE(-1)	-0.132504	0.087700	-1.510866	0.1401
C	0.457009	0.303188	1.507346	0.1410

R-squared	0.062915	Mean dependent var	0.007614
Adjusted R-squared	0.035353	S.D. dependent var	0.358889
S.E. of regression	0.352488	Akaike info criterion	0.806351
Sum squared resid	4.224418	Schwarz criterion	0.894324
Log likelihood	-12.51431	Hannan-Quinn criter.	0.837056
F-statistic	2.282716	Durbin-Watson stat	2.240675
Prob(F-statistic)	0.140062		

source: Agaton and Karl (2018)

Chapter 5

Real Options Analysis of Renewable Energy Investment Scenarios in the Philippines

Abstract - This chapter is based on a paper published in *Renewable Energy and Sustainable Development*[6] journal. With the continuously rising energy demand and much dependence on imported fossil fuels, the Philippines is developing more sustainable sources of energy. Renewable energy seems to be a better alternative solution to meet the country's energy supply and security concerns. Despite its huge potential, investment in renewable energy sources is challenged with competitive prices of fossil fuels, high start-up, and lower feed-in-tariff rates for renewables. This chapter addresses these problems by analyzing energy investment scenarios in the Philippines using real options approach. This compares the attractiveness of investing in renewable energy over continuing to use coal for electricity generation under uncertainties in coal prices, investments cost, electricity prices, growth of investment in renewables, and imposing carbon tax for using fossil fuels.

---

[6]Agaton (2017b) available at `https://doi.org/10.1186/s13705-017-0143-y` licensed under *Creative Commons Attribution-NonCommercial 4.0*

5.1   Introduction

Increasing environmental concerns and depleting fossil fuels have caused many countries
to find cleaner and more sustainable sources of energy. Currently, renewable energy
sources (RES) supply 12.65% of the total world energy demand in 2016 which includes
wind, solar, hydropower, biomass, geothermal, and ocean energies (EIA (2017b)). In the
recent years, new investments in renewable energy have grown from US$1043.8B (2007-
2011) to US$1321.9 (2012-2016) with a geographic shift from the Asia-Pacific region
(BNEF (2016); FS-UNEP (2017)). In the Philippines, renewable energy accounts to
25% of the energy generation mix, mostly from geothermal (13%) and hydropower (10%)
(DOE (2016a)). The country is aiming to increase this percent share to 60% in 2030 by
investing and developing localized renewable sources at 4% annual growth rate (DOE
(2012)). According to International Renewable Energy Agency (IRENA), the country's
topography and geographic location makes a good potential for renewable energy with
170GW from ocean, 76.6GW from wind, 4GW from geothermal, 500MW from biomass,
and 5kWh/m2/day from solar energy (IRENA (2017)). Despite its potential, the
country's 60% renewable energy goal seems unachievable as the growth in electricity
demand increases faster than investment and generation from RES. Meanwhile, the
country is burdened by heavy dependence on imported fossil fuels particularly coal
and oil. As more power plants are needed due to closing old coal plants and rising
electricity demand, renewable energy seems to be the long-term solution to address
the country's problem on energy security and sustainability. However, investment
in renewable energy sources is challenged by competitive prices of fossil fuels, high
investment cost, and lower feed-in tariff (FiT) rates for renewables. These serve as an
impetus to evaluate the comparative attractiveness of renewable energy over coal for
electricity generation in the Philippines.

This study presents a general framework of investment decision-making for shifting technologies from coal to renewable sources that can be applied to developing countries. By taking the case of the Philippines, this study applies the real options approach (ROA) to analyze various investment scenarios. Traditionally, the discounted cash flow (DCF) or net present value (NPV) techniques are mostly used in valuating investment projects. These methods, however, do not cover highly volatile and uncertain investments because they assume a definite cash flow. This assumption makes DCF and NPV underestimate the investment opportunities leading to poor policy and decision-making process particularly to energy generation projects. Further, these approaches do not allow an investor to define the optimal time to invest or to estimate the true value of project uncertainties (Santos et al. (2014)). ROA overcomes this limitation as it combines risk and uncertainty with flexibility of investment as a potential positive factor, which gives additional value to the project (Brach (2003)). This approach evaluates investment projects by considering the investor's flexibility to delay or postpone his/her decision to a more favorable situation (Kumbaroglu et al. (2008)). These ROA characteristics are highlighted in this paper as the decision-making process to invest in RES is evaluated in every investment period (annually) using dynamic optimization under various uncertainties.

Recent studies employ ROA to analyze investment decisions, specifically renewable energy, including: Zhang et al. (2016) on investment in solar photovoltaic (PV) power generation in China by considering uncertainties in unit generating capacity, market price of electricity, CO_2 price, and subsidy; Kim et al. (2017) on analyzing renewable energy investment in Indonesia with uncertainties in tariff, energy production, Certified Emission Reduction price, and operations and maintenance (O& M) cost; Kitzing et al. (2017) on analyzing offshore wind energy investments in the Baltic under different support schemes as FiT, feed-in premiums, and tradable green certificates; Tian et al. (2016)on valuating PV power generation under carbon market linkage in carbon price, electricity price, and subsidy uncertainty; and Ritzenhofen and Spinler (2016) on assessing the impact of FiT on renewable energy investments under regulatory uncertainty. This research contributes to existing literatures by presenting a multi-period investment coupled with uncertainties in coal prices, cost of renewable technologies, growth of renewable energy investment, FiT price of renewables, and externality for using coal.

The main goal of this paper is to analyze investment scenarios that make renewable energy a better option than continuing to use coal for electricity generation. Specifically, this study employs ROA to evaluate the (1) maximized option value of either continuing to use coal or investing in renewables, (2) value of waiting or delaying to invest in renewables, and (3) optimal timing of investment characterized by the trigger price of coal for shifting technologies from coal to renewables. Sensitivity analyses are done to investigate how the above-mentioned uncertainties affect the optimal investment strategies.

5.2   Methodology

The proposed real options methodology is divided into two subsections. The first subsection describes dynamic optimization to calculate the maximized value of investment and identify the optimal timing of investment. The second stage includes the sensitivity analyses with respect to growth rate of renewable energy investment, prices of renewable energy, investment costs, and CO_2 prices.

5.2.1   Real options model

Consider a renewable energy project with lifetime T_R, which can be irreversibly initiated in three installment periods τ, $\tau+5$, and $\tau+10$ with investment costs I_{R_0}, I_{R_5}, and $I_{R_{10}}$. Assume that the project construction can be finished instantaneously and operated in full load after project completion. If renewable energy project starts in period t, the total net present value of the project NPV_R can be represented by Equation 5.1.

$$NPV_R = NPV_{R_0} + NPV_{R_5} + NPV_{R_{10}} = \sum_{r=0,5,10} \left\{ \sum_{t=\tau+r}^{T_R+r} \rho^t PV_{R,t} - \{1+\phi_r\} I_{R,r} \right\} \quad (5.1)$$

where r is the installment periods of renewable energy investment, ϕ_r is the growth of renewable energy investment cost, and τ is the period where investor decides to invest in renewable (see Appendix Table 5.1 for the list of all variables and estimation parameters).

The yearly cash flow $PV_{R,t}$ of renewable energy project comprises of returns from selling electricity from RE, O&M cost C_R.

$$PV_R = \pi_R = P_{ER}Q_R - C_{R,r} \quad (5.2)$$

On the other hand, there exists a power plant generated with coal. The net present value of yearly cash flow from coal depends on the returns from selling electricity from coal, O&M cost C_C, stochastic cost of fuel $P_{C,t}$, and CO_2 price $C_{C_{CO_2}}$ as given in Equation 5.3

$$NPV_C = \sum_{t=0}^{\tau} PV_{C,t} = \sum_{t=0}^{\tau} \rho^t \pi_{C,t} = \sum_{t=0}^{\tau} \rho^t \left\{ P_{EC}Q_R - P_{C,t}Q_C - C_C - C_{C_{CO_2}} \right\} \quad (5.3)$$

where ρ is the social discount factor, P_{EC} and P_{ER} are the prices of electricity from coal and renewable, Q_R is the quantity of electricity generated from coal/renewable, and Q_C is the quantity of coal needed to generate Q_R.

Following previous literatures, this research assumes that the price of coal is stochastic and follows GBM (Dixit and Pindyck (1994); Tietjen et al. (2016); Wang and Du (2016); Xian et al. (2015); Detert and Kotani (2013)). The current price of P_C depends on its previous price, and the drift and variance rates of time series of coal prices as shown in Equation 5.4

$$P_{C,t} = P_{C,t-1} + \alpha P_{C,t-1} + \sigma P_{C,t-1} \varepsilon_{t-1} \tag{5.4}$$

with α and σ are the GBM rate of drift and variance of coal prices, and ε a standard normally distributed random number such that $\varepsilon_t \ N(0,1)$.

The parameters α and σ are approximated using ADF test from time series of coal prices (Insley (2002)). The estimates obtained in ADF test: $\alpha = 0.032027$ and $\sigma = 0.2494$ are used to generate a matrix of random numbers that represent possible prices of coal from initial values of zero to US\$200 at every investment period from zero to T (see Appendix Table 5.2 for ADF unit root test result). These values are then used to calculate the present values of electricity generation from coal for each period.

Using Monte Carlo simulation, the expected NPV for generating electricity from coal is estimated by calculating the $NPV_{C,t}$ in Equation 5.3 and repeating the process for a sufficiently large number $J = 10000$ times. Expected net present value is calculated by taking the average of NPV_C for every initial price of coal $P_{C,0}$ as shown in Equation 5.5.

$$\mathbb{E}\left\{NPV_{C,j} \mid P_{C,0}\right\} \approx \frac{1}{J}\sum_{j=1}^{J}NPV_{C,j} \approx \mathbb{E}\left\{NPV_C \mid P_{C,0}\right\} \tag{5.5}$$

The next exercise in identifying the optimal timing and associated trigger price of coal for shifting technologies is done with dynamic optimization as shown in Equation 6.

$$\max_{0 \leq \tau < T+1} \mathbb{E}\left\{\sum_{t=0}^{\tau}\rho^t\pi_{C,t} + \sum_{t=\tau}^{T}\rho^t\pi_{C,t}\left\{1 - \mathbb{I}_{\{\tau \leq T\}}\right\} + \left\{NPV_R + NPV_C\{\mathbb{I}_{\{\tau \leq T\}}\}\right\}\right\} \tag{5.6}$$

where $\mathbb{I}_{\{\tau \leq T\}}$ is an indicator equal to 1 if switching to renewable energy, otherwise, equal to 0. This model describes an investor who is given a specific period T to decide whether to continue generating electricity from coal or invest in renewable energy. In this model, $\sum_{t=0}^{\tau}\rho^t\pi_{C,t}$ accounts to the net present value of using coal from initial period $T = 0$ until τ when the investor makes the decision. If the investor chooses not to invest ($\mathbb{I}_{\{\tau \leq T\}} = 0$), he/she incurs a net present value of $\sum_{t=\tau}^{T}\rho^t\pi_{C,t}$ from period τ until the end of the coal plant's lifetime. If the investor chooses to invest ($\mathbb{I}_{\{\tau \leq T\}} = 1$), he/she incurs a net present value of NPV_R from successive (three-period) investment in renewables plus NPV_C, as generation from coal will continue at a lower quantity because other electricity will be generated from renewables.

From Equation 5.6, the investor's problem is to choose the optimal timing of investment τ, to maximize the expected net present value of investment. The problem is solved backwards using dynamic programming from the terminal period for each price of coal $P_{C,t}$ as shown in Equation 5.7

$$V_t(P_{C,t}) = \max\left\{\pi_{C,t-1} + V_t\left(P_{C,t-1}\right), NPV_R + NPV_C\right\} \tag{5.7}$$

with V_t as the option value of investment at coal price $P_{C,t}$.

The optimal timing of investment $\widetilde{P_C}$ is characterized by the minimum price of coal so that switching to renewable energy is optimal as shown in Equation 5.8 (Detert and Kotani (2013); Davis and Cairns (2012)).

$$\widetilde{P_C} = min\,\{P_{C,t} \mid V_0(P_{C,t}) = V_{T_A}(P_{C,t})\} \tag{5.8}$$

Finally, investment strategy is described by a decision to invest when $\widetilde{P_C} \leq P_C$, otherwise, investment can be delayed in later periods until $\widetilde{P_C} = P_C$.

5.2.2   Parameter Estimation and Investment Scenarios

The following scenarios describe various environments that affect investment decisions in renewable energy in the Philippines. Sensitivity of investment values and optimal timing are analyzed with respect to growth rate of renewable energy investment, price of electricity from renewable energy, investment cost, and carbon prices.

The first scenario is the BAU case which describes the current renewable energy investment scenario in the country. To estimate a suitable set of parameters in this scenario, secondary data from the DOE and EIA are used (DOE (2017)); EIA (2017c)) A 30-year period of average annual coal prices from 1987-2016 is used to run the ADF test described in Equation 4. The ADF test result (see Appendix - Table 5.2) implies that the null hypothesis that p_t has a unit root cannot be rejected at all significant levels, hence, coal prices conform with GBM. From this test, the estimated GBM parameters are $\alpha = 0.032027$ and $\sigma = 0.249409$, and are used to approximate stochastic prices of coal for each investment period. The social discount rate is set to 7.5%. From Equation 3, $C_{C_{CO_2}}$ is set to zero as there are no existing carbon prices in the Philippines at present. The growth rate of renewable energy investment is set to 2% per annum. This is equivalent to 470GWh of electricity generation from renewables. From this value, the investment cost and operations and maintenance cost for renewables are estimated, as well as the costs and quantity of coal needed to generate this amount. The prices of electricity, $P_{ER} = P_{EC}$ =US\$182.2/MWh are set equal to the current domestic electricity price, constant during the entire investment period, and independent of the domestic demand. Assumptions indicate that renewable energy sources can generate electricity at an annual average of Q_R all throughout its lifetime; there are no technological innovations that affect energy efficiency and overnight costs of renewables; and stochastic prices of coal are independent of the demand for renewable energy.

The second scenario describes a situation of an accelerated growth rate of renewable energy investment from the current 2% to 4%, 6%, and 8%. Meanwhile, the third scenario analyzes the effect of prices of electricity from renewable energy by increasing the current FiT rates to proposed rates. Three prices are set: US\$182.2/MWh at the BAU case, US\$160/MWh which is 10% lower than the BAU case, and US\$200/MWh which is 10% higher. The third scenario describes a situation of a decline in investment costs for renewable energies by 5%, 10%, and 15% respectively. The last scenario introduces a government policy of introducing carbon tax for electricity generation from coal. The carbon tax is set to US\$ 0.504/MWh (see Appendix Table 5.3 for all estimation parameters used in the dynamic optimization).

5.3   Results and Discussion

5.3.1   Business as usual scenario

The dynamic optimization process described in the previous section results in three significant values. First is the option value which is equal to the maximized value of either investing in renewables or continuing to use coal. Second is the value of waiting as described by the vertical distance between option value curves: initial period (dotted) and terminal period (bold) of investment. This value approximates the gains of an investor if investment is delayed or postponed to some period. The last estimated value is the the optimal timing of investment denoted by the trigger prices of coal for shifting electricity source from coal to renewables. This trigger price is illustrated as the intersection of the two option value curves, and indicates the threshold where the value of waiting is zero and that an investor has no benefit to delay the investment to renewables.

Figure 5.1 shows the dynamics of option values at different prices of coal in the business as usual scenario. The first point of interest is the positive option values. It indicates that investment in renewable energy incurs positive returns at the current energy situation in the Philippines. This contradicts with the result of Detert and Kotani (2013) where the optimization yields negative option values describing a government controlled, operated, and subsidized energy regime. The next point of interest is option value curves sloping downward. This indicates that option values decrease with increasing cost for input fuel. At certain point on the curves, the option values become constant. These are the prices of coal where investment in renewable is a better option than continuing to use coal for electricity generation. The positive values

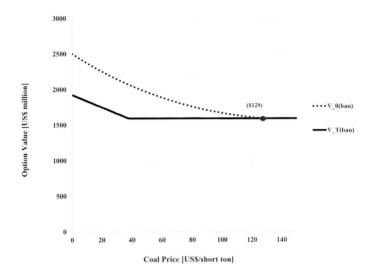

Figure 5.1: Option values at the business as usual scenario

Note: V_0(bau) -option values for shifting energy source for every price of coal at the initial period of investment; V_T(bau) -option values for shifting energy source for every price of coal at the terminal period of investment.

source: Agaton (2017b)

further indicate positive NPV_R for investing in renewable. In this scenario, the result shows that the trigger price of coal for shifting technologies is US$ 129/short ton. This trigger price is higher than the current price of coal US$93.13/short ton (year 2016), and implies that delaying investment in renewables is a better option. However, at the current coal price, the value of waiting to invest is -US$105.4 million. This negative value indicates possible losses incurred from delaying investment in renewables.

5.3.2   Growth rate of renewable energy investment scenario

This scenario describes an accelerated growth of investment in renewable energy sources. While the country is aiming to increase the current share of energy generation from renewables, from 25% to 60% by 2030 at 4% annual growth rate (DOE (2012)), this goal seems unattainable as the country's electricity demand is increasing at a faster rate than renewable investments(DOE (2017)). This scenario examines how changing the rate of growth in renewable energy investment affects the option values and trigger prices.

The results of dynamic optimization at various growth rates are shown in Figure 5.2. It can be observed that option value curves shift upwards. This implies that increasing investment in renewables incurs higher returns from economies of scale. Doubling of wind farms could result in price reductions as the costs can be spread over large production of electricity (Dai et al. (2016); Qiu and Anadon (2012); Junginger et al. (2005)). It can be noticed that the trigger prices of coal have also decreased from US$129/short ton in the BAU scenario, to US$120, US$113, and US$105 at 4%, 6%, and 8% growth rates. Finally, the value of waiting to invest varies from -US$105.4M at BAU scenario, to -US$139.5M at 4% growth, -US$146M at 6%, and -US$153.7M at 8% growth rates. These results suggest that accelerating the current growth rate from

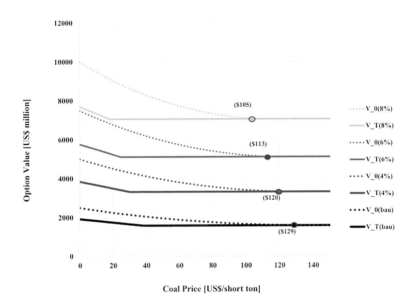

Figure 5.2: Option values at different rates of renewable energy investment
Note: V_0(bau) -option values for shifting energy source for every price of coal at the initial period of investment for the BAU(2% growth); V_T(bau) -option values at the terminal period of investment for the BAU(2% growth); V_0(4%) -option values at the initial period with 4% growth of RE investment; V_T(4%) -option values at the terminal period with 4% growth of RE investment; V_0(6%) -option values at the initial period with 6% growth of RE investment; V_T(6%) -option values at the terminal period with 6% growth of RE investment; V_0(8%) -option values at the initial period with 8% growth of RE investment; V_T(8%) -option values at the terminal period with 8% growth of RE investment.
source: Agaton (2017b)

business as usual prevents potential losses from waiting to invest in renewables.

5.3.3   Price of electricity from renewable energy

In this scenario, the effect of changing electricity prices from renewables on option
values and trigger prices is analyzed. Currently, the Philippines is one of the countries
with the highest electricity rates in the Asia-Pacific region. Compared with neighboring
countries including Thailand, Malaysia, South Korea, Taiwan, and Indonesia, the prices
are lower as the government subsidized the cost through fuel subsidy, cash grants,
additional debt, and deferred expenditures. In the Philippines, electricity prices are
higher due to no government subsidy, fully cost-reflective, imported fuel-dependent,
and heavy taxes across the supply chain (Fernandez (2015); IEC (2016)). By changing
the value broadly, this scenario presents how potential government actions regarding
electricity prices affect investment conditions in renewable energy.

Figure 5.3 illustrates the optimization outcomes with varying electricity prices. The
result shows an upward shift of option values at higher electricity prices. This result is
expected as higher price increases the revenues and the net present value of electricity
generation from renewable energy. On the other hand, the result shows the inverse
relationship of electricity prices and trigger prices from US$129/MWh in BAU to
US$100/MWh at 10% higher and US$159/MWh at 10% lower electricity price. The
values of waiting to invest also show a similar trend from -US$105.4M at BAU to
-US$25.9M at higher and -US$241.6M at lower electricity price. This implies that
setting the price of electricity generated from renewables higher than current tariff
provides a better environment for renewable energy investments. Nevertheless, this
study also considers the possibility that extensive electricity generation from renewable
energy sources has significant impact on the electricity prices as stated in previous

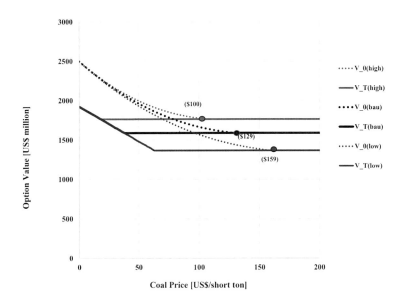

Figure 5.3: Option values at various electricity prices from renewable sources
Note: V_0(bau) -option values for shifting energy source for every price of coal at the initial period of investment for the BAU; V_T(bau) -option values at the terminal period of investment for the BAU; V_0(low) -option values at the initial period with lower electricity price; V_T(low) -option values at the terminal period with lower electricity price; V_0(high) -option values at the initial period with lower electricity price; V_T(high) -option values at the terminal period with lower electricity price.
source: Agaton (2017b)

literatures(Ketterer (2014); Jonsson et al. (2010); Gelabert et al. (2011)).

5.3.4   Investment cost scenario

This scenario describes how decline in overnight cost affects investment in renewables. In the recent years, growth in renewable energy investments is driven by several factors including the improving cost-competitiveness of renewable technologies, policy initiatives, better access to financing, growing demand for energy, and energy security and environmental concerns (FS-UNEP (2017); REN21 (2016)). This scenario focuses on the effect of renewable energy cost on investment option values and trigger prices of coal for shifting technologies.

Figure 5.4 shows the dynamics of option values at various investment cost scenarios. The result shows an upward shift in the option value curves. This outcome is evident as lower investment cost incurs higher net present value for renewable energy, leading to higher option values. The trigger prices decrease from US\$129 in BAU to US\$124, US\$119, and US\$114 at 5%, 10%, and 10% cost reduction. The value of waiting also decreases from -US\$105.4M in BAU to -US\$86.5M, -US\$68.6M, and -US\$52.2M respectively. This result confirms the rapid growth in investment as caused by the sharp decline in renewable technology costs.

5.3.5   Externality scenario

The last scenario discusses the effect of carbon prices for electricity generation from coal. Currently, there are no carbon prices in the Philippines. This study evaluates the effect of imposing carbon tax as proposed in previous literatures(Meller and Marquardt (2013); Burtraw and Krupnick (2012); Cabalu et al. (2015)). As shown in Figure

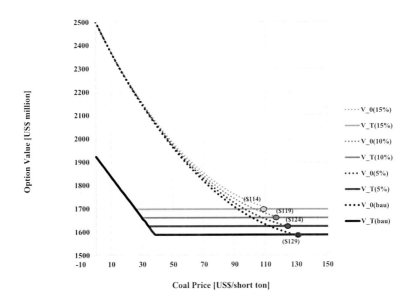

Figure 5.4: Option values at different decline of renewable investment cost
Note: V_0(bau) -option values for shifting energy source for every price of coal at the initial period of investment for the BAU(no decline); V_T(bau) -option values at the terminal period of investment for the BAU(no decline); V_0(5%) -option values at the initial period with 5% decline in RE cost; V_T(5%) -option values at the terminal period with 5% decline in RE cost; V_0(10%) -option values at the initial period with 10% decline in RE cost; V_T(10%) -option values at the terminal period with 10% decline in RE cost; V_0(15%) -option values at the initial period with 15% decline in RE cost; V_T(15%) -option values at the terminal period with 15% decline in RE cost.
source: Agaton (2017b)

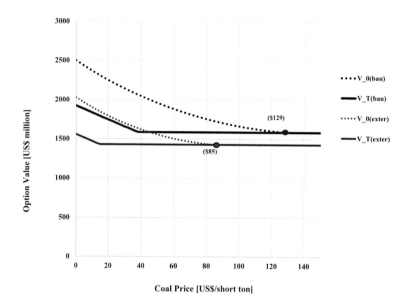

Figure 5.5: Option values with externality cost for using coal
Note: V_0(bau) -option values for shifting energy source for every price of coal at the initial period of investment for the BAU; V_T(bau) -option values at the terminal period of investment for the BAU; V_0(exter) -option values at the initial period with externality tax for using coal; V_T(exter) -option values at the terminal period with externality tax for using coal.
source: Agaton (2017b)

5.5, the option values and trigger prices decrease with the addition of externality cost. This result is anticipated as additional cost decreases the value of electricity generation from coal. It can also be noted that the trigger price is lower than the current price of coal equal to US\$93/short ton (year 2006). This implies that investing in renewable is a better option than continuing to use coal if carbon tax is imposed. Furthermore, with carbon tax, the demand for carbon-intensive inputs, including coal and oil, will decrease, while less carbon- and carbon free energy inputs eventually increase. This finally supports the research aim of analyzing renewables as a cleaner and more sustainable source of energy and a better alternative to coal.

5.4 Conclusion

This study presented various investment scenarios that represent energy switching decisions that apply to developing countries. By taking the case of the Philippines, this study employed real options approach to evaluate the maximized option values of investing in renewables, value of deleying investment, and trigger prices of coal for shifting technologies from coal to renewable sources. While numerous studies applied this approach to analyze renewable energy investments, this study expanded the existing body of research by considering a multi-period investment and taking account of uncertainties in input fuel prices, renewable technology cost, growth of investment in renewables, and externality cost for using coal.

The analyses conclude that renewable energy is a better option than continuing to use coal for electricity generation in the Philippines. Delaying the investment in renewables may lead to possible welfare losses. Shifting from fossil-based to renewable sources is very timely as the costs of renewable technologies have decreased immensely throughout the years and expected to continuously fall. To support investments in renewable energy, the government must set higher FiT rates than business as usual, and impose carbon tax for using carbon-intensive fuels. Further, the growth in investment in renewables should be increased to meet the country's goal of 60% energy generation from renewable sources and decrease its dependence on imported fossil fuels.

While this study compared coal and renewables, particularly wind energy, for electricity generation, future studies may also analyze other sustainable energy sources including hydropower, solar, geothermal, biomass, tidal/ocean, and other technologies designed to improve energy efficiency. Lastly, environmental uncertainty such as climate variability and weather disturbances, that affects energy generation may also be included to further capture investment scenarios relevant to climate change policy.

5.5 Appendix

Table 5.1: List of Variables and Estimation Parameters

V_t	Option value of investment at each price of coal at each period t, US$
π_R	Profit for investing in renewable energy, US$
$\pi_{C,t}$	Profit of using coal for electricity generation, US$
NPV_C	Net present value of using coal for electricity generation, US$
NPV_R	Net present value of investing in renewable, US$
PV_R	Present value of renewable energy, US$
P_{ER}	Price of electricity from renewable energy, US$/MWh
P_{EC}	Price of electricity from coal, US$/MWh
$P_{c,t}$	Stochastic price of coal, US$/short ton
Q_R	Quantity of electricity from renewable/coal, MWh
Q_c	Quantity of coal needed to produce Q_R, short ton
C_c	Annual marginal operations and maintenance cost for electricity generation using coal, US$
C_R	Annual marginal operations and maintenance cost for electricity generation from renewable, US$
$C_{C\_CO_2}$	CO_2 cost from coal, US$
I_R	Investment cost for renewable energy, US$
r	Installment periods of renewable energy investment
T_R	Lifetime of electricity generation from renewable energy, years
T	Total period of investment, years
τ	Period where investor decides to invest in renewable
ρ	Discount factor
α	GBM rate of drift of coal prices
σ	GBM variance of coal prices
φ	Growth of renewable energy investment cost
$\mathbb{1}_{\tau \leq T}$	Indicator equal to 1 if switching to renewable or energy is made, otherwise, equal to 0
J	Number of times for Monte Carlo simulation process
V_0	Option value at the initial period of investment
V_T	Option value at the terminal period of investment

source: Agaton (2017b)

Table 5.2: ADF Unit Root Test result for Coal Prices (1991-2016)

Null Hypothesis: LNPRICE has a unit root
Exogenous: Constant
Lag Length: 3 (Automatic - based on SIC, maxlag=7)

		t-Statistic	Prob.*
Augmented Dickey-Fuller test statistic		-1.119480	0.6924
Test critical values:	1% level	-3.711457	
	5% level	-2.981038	
	10% level	-2.629906	

*MacKinnon (1996) one-sided p-values.

Augmented Dickey-Fuller Test Equation
Dependent Variable: D(LNPRICE)
Method: Least Squares
Date: 09/11/17   Time: 16:48
Sample (adjusted): 1991 2016
Included observations: 26 after adjustments

Variable	Coefficient	Std. Error	t-Statistic	Prob.
LNPRICE(-1)	-0.103245	0.092226	-1.119480	0.2756
D(LNPRICE(-1))	0.226008	0.205954	1.097368	0.2849
D(LNPRICE(-2))	-0.377858	0.199774	-1.891430	0.0724
D(LNPRICE(-3))	0.557382	0.213207	2.614279	0.0162
C	0.414604	0.356738	1.162208	0.2582

R-squared	0.359108	Mean dependent var	0.032027
Adjusted R-squared	0.237033	S.D. dependent var	0.249409
S.E. of regression	0.217853	Akaike info criterion	-0.038947
Sum squared resid	0.996662	Schwarz criterion	0.202994
Log likelihood	5.506316	Hannan-Quinn criter.	0.030723
F-statistic	2.941702	Durbin-Watson stat	1.563558
Prob(F-statistic)	0.044650		

source: Agaton (2017b)

Table 5.3: Summary of Estimation Parameters for Dynamic Optimization

Parameter	Value	Unit	Description
alpha	0.032027		estimated myu value of GBM unit root test of coal prices
sigma	0.249409		standard deviation of GBM unit root test of coal prices
rho	0.91		discount factor
Pmin	0	US$/short ton	base-level for price of coal
Pmax	200	US$/short ton	maximum limit for price of coal
Pstep	1	US$/short ton	the value between each price node
P_e	182.2	US$/MWh	price of electricity
Q_e	1409.4	GWh	average annual electricity generated from coal
Q_e1	939.6	GWh	annual electricity generated from coal after first installment of renewables
Q_e2	469.8	GWh	annual electricity generated from coal after second installment of renewables
Q_c	862241	short ton	average annual quantity of coal used to generate Q_e
Q_c1	574827	short ton	average annual quantity of coal used to generate Q_e1
Q_c2	287414	short ton	average annual quantity of coal used to generate Q_e2
C_c	28.2M	US$	annual O&M cost to generate Q_e from coal
C_c1	18.8M	US$	annual O&M cost to generate Q_e1 from coal
C_c2	9.4M	US$	annual O&M cost to generate Q_e2 from coal
Q_w1	469.8	GWh	average annual electricity generated from renewables after the first installment period
Q_w2	939.6	GWh	average annual electricity generated from renewables after the second installment period
Q_w3	1409.4	GWh	average annual electricity generated from renewables after the third installment period
C_w	6.2M	US$	annual O&M cost for Q_w1
I_w1	561M	US$	investment cost of renewables on the first installment
I_w2	323M	US$	investment cost of renewables on the first installment
I_w3	298M	US$	investment cost of renewables on the first installment
LL	30	years	time horizon for dynamic optimization problem
T_c	15	years	number of periods for NPV of coal
T_r	30	years	time horizon for renewable energy generation
C_CO2	0.504	US$	Carbon cost
160, 182.2(BAU), 200		US$/MWh	Prices of electricity from renewables
2%(BAU), 4%, 6%, 8%			Growth rate of renewable energy investment
0%(BAU), 5%, 10%, 15%			Decline in renewable technology cost

source: Agaton (2017b)

Chapter 6

Conclusion and Recommendation

With the increasing concern on the environmental effects of burning fossil fuels, both developed and developing countries are exploring plausible solutions including investment in cleaner and more sustainable sources of energy. This research addressed this energy transition problem by taking the case of the Philippines. Using real options approach under uncertainty, this research presented a general framework of technology switch from fossil fuel-based to alternative energy sources that applies to developing countries, particularly to those that are highly dependent on imported fossil fuels. Although numerous studies used real options approach to evaluate energy investments, this research expanded the existing body of research by including the analysis of various alternative energy sources, uncertainty in fuel prices, sensitivity in electricity price, negative externality, social discount rate, RE investment cost, growth of RE investment, multi-period investment, and the risk of possible nuclear accident. Using dynamic optimization, this research evaluated the option values of investments, value of waiting to invest, and the trigger prices of fuel for shifting technologies to alternative energy. This dissertation divided the analyses of results into four chapters.

Chapter 2 focused on investment with various renewable energy source namely: wind, solar PV, geothermal, and hydropower. This chapter evaluated whether investment on these renewable sources is better than continuing to use coal for electricity generation in the Philippines. In this chapter, ROA is done under uncertainty in coal prices. The results highlighted the attractiveness of investing in renewable energy than coal. Among the renewable energy sources, geothermal seems to be the most attractive followed by hydropower, wind, and solar. Although geothermal and hydropower are already established sources of renewable energy in the Philippines, wind and solar PV are still promising as the investment costs on these alternatives continuously drop in the recent years (FS-UNEP (2017)). The trigger prices of coal for shifting technologies for all renewable sources were relatively higher than the current (2016) price of coal. This suggests that the government must increase the coal prices by imposing import tax for the coal to be less competitive and attract more investments in renewables. On the contrary, negative values of option to wait show that delaying or waiting to invest incurs losses. In the sensitivity analysis, results implied a better opportunity to invest earlier on renewables with more deterministic prices of coal. In contrast, higher volatility in coal prices suggested a longer waiting to shift to renewables to avoid investment risks. Further, decreasing the social discount rate from its current rate adds benefit for power producers to shift energy source from coal to renewables.

Chapter 3 compared the attractiveness of investing in alternative sources (renewable energy or nuclear energy) over coal. In this chapter, the probability of having nuclear accident is incorporated in the real options model with uncertainty in coal prices as proposed in Chapter 2. This also considered the negative externality of using various sources of energy. Initial results revealed good prospects for investing in both alternatives. Although results showed, that it is more optimal to wait or delay the investment on these alternatives, the negative values of waiting suggest the otherwise. The inclusion of nuclear accident risk favors investment in renewable energy over nuclear. Further, inclusion of externality costs favors investment on both alternatives over coal. Being nonrenewable and exhaustible, the concerns on coal's limited supply, price volatility, national security problems, and the environmental effects associated with its continued use serve as an impetus of finding better and more sustainable sources of energy. Despite the accident risk, nuclear energy still seemed to be attractive in the Philippines. Since the country is still skeptical about the radiation and health risks of nuclear, investment on this alternative should only serve as a transition technology until renewable energy becomes more competitive than coal. To make renewable energy more attractive than coal and nuclear, the government requires some intervention by subsidizing electricity from renewable energy, imposing carbon or environmental tax for using coal, or setting a high insurance premium for nuclear investment.

Chapter 4 evaluated various investment environments for substituting diesel power plant with renewable energy for electricity generation in the island of Palawan. The rationale of this chapter was to apply the real options model with a medium scale investment, focusing on an island not connected to the national grid. This chapter analyzed the effect of varying local electricity prices and externality on renewable energy investment using real options under diesel price uncertainty. The result highlighted the attractiveness of switching technologies from bunker-fired power plant to solar PV. Similar with the results in previous chapters, the negative value of option to wait implied that investing in solar PV earlier is more optimal than waiting or delaying to invest. In the electricity price scenario, setting higher electricity favored the use of diesel. This suggested the local government policy on decreasing the electricity price in order to attract investors in renewable energy and push the local power producers to shift technologies from diesel to renewable. Imposing negative externality for using diesel favored investment in renewable energy. The result also presented a threshold of externality cost that made the trigger price equal to the current price of diesel. In order for local producers to switch technologies, a government policy on externality tax for using diesel must be set equal to this threshold value. With the booming trend in solar PV investment in the country, this would set the opportunity to invest in renewables in order to address the islands problem on energy security, regular electricity roll-out, and environmental effects of burning diesel.

Chapter 5 expanded the proposed real options model from previous chapters by introducing a multi-period investment and taking account of uncertainties in input fuel prices, renewable technology cost, growth of investment in renewables, sensitivity in electricity prices, and externality cost for using coal. In line with previous chapters, the analyses in this chapter conclude that renewable energy is a better option than continuing to use coal and that delaying the investment in renewables may lead to possible welfare losses. Shifting from fossil-based to renewable sources is very timely as the costs of renewable technologies have decreased immensely throughout the years and expected to continuously fall. To support investments in renewable energy, the government must set higher FiT rates than business as usual and impose carbon tax for using carbon-intensive fuels. Further, the growth in investment in renewables should be increased to meet the country's goal of 60% energy generation from renewable sources and decrease its dependence on imported fossil fuels.

To develop the real options model, this research made several simplifying assumptions leading to various limitations in the analyses. The major limitation of this study was the availability and reliability of data as the Philippines had limited data on energy investments particularly nuclear. In this case, several international sources of data were used including those from Nuclear Energy Agency and Energy Information Administration. Another was the rough approximation of costs associated with investments, externality, nuclear damage, and other parameters to estimate the option values. While the assumptions used in this study are sufficient enough for the main objective of providing qualitative guidance and general scenario of energy investment, it must be noted that thorough parameter estimations requiring calculations with more tailored numerical methods are necessary in real decision-making.

Extensions of the framework presented in this dissertation could be used for further applications. While this study compared fossil-based (coal and diesel) with alternative sources (nuclear and renewables, particularly wind, hydro, solar PV, and geothermal) future studies may also analyze other sustainable energy sources including tidal, ocean/wave, biomass, hydrogen, and other technologies designed to improve energy efficiency including hybrid energy systems. Hybrid systems, consisting of two or more energy sources, could provide increased system efficiency, balance energy supply, and more reliable and continuous source of energy.

This study assumed that prices of fossil fuels are stochastic, follow GBM with a drift, and increasing in the long run. With fossil fuels being exhaustible resources, the current demand path and competetion accelerates the upward trend in prices with higher volatility. Nevertheless, this study acknowledges that global investments in renewables may drive down the demand and prices of fossil fuels in the long run. This trend in fossil fuel demands and prices should also be accounted for. Another limitation is that the current study focuses only on the financial valuation of energy investments. In real project valuation and decision-making process, there are also several factors considered including economic impacts on employment and local economy; environmental impacts on landscape, wildlife, habitat loss, air and water pollution; and socio-technical factors including user's energy demand, usage patterns, system sizing, and avaliability of renewable resources. Future studies could incorporate these factors to make the current ROA model more robust and valuable not only to investors but to project evaluators and policy makers as well. Finally, environmental uncertainty, such as climate variability and weather disturbances, that affects energy generation may also be included to further capture investment scenarios relevant to climate change policy. Yet, there would be plentiful studies that could be derived from this research. The researcher is optimistic that this contribution becomes one-step forward for further analysis for investment in more sustainable sources of energy.

BIBLIOGRAPHY

Abadie, L. M. and Chamorro, J. M. (2014). Valuation of wind energy projects: a real options approach. *Energies*, 7:3218–3255. https://doi.org/10.3390/en7053218.

Agaton, C. (2017a). Coal, renewable, or nuclear? a real options approach to energy investments in the philippines. *International Journal of Sustainable Energy and Environmental Research*, 2:50–61. https://doi.org/10.18488/journal.13.2017.62.50.62.

Agaton, C. (2018a). *Transition Towards 100% Renewable Energy: Selected Papers from the World Renewable Energy Congress WREC 2017*, chapter To import coal or invest in renewable: a real optios approach to energy investments in the Philippines, pages 1–10. Springer, Chem. https://doi.org/10.1007/978-3-319-69844-1_1.

Agaton, C. B. (2017b). Real options analysis of renewable energy investment scenarios in the philippines. *Renewable Energy and Sustainable Development*, 3(3):284–292. https://doi.org/10.21622/RESD.2017.03.3.284.

Agaton, C. B. (2018b). *A real options approach to renewable and nuclear energy investments in the Philippines*. Ruhr University of Bochum.

Agaton, C. B. (2018c). Use coal or invest in renewables: a real options analysis of energy investments in the philippines. *Renewables: Wind, Water, and Solar*, 5:1–8. https://doi.org/10.1186/s40807-018-0047-2.

Agaton, C. B. and Karl, H. (2018). A real options approach to renewable electricity generation in the philippines. *Energy, Sustainability and Society*, 8:1–9. https://doi.org/10.1186/s13705-017-0143-y.

Akinyele, D. and Rayudu, R. (2016). Strategy for developing energy systems for remote communities: Insights to best practices and sustainability. *Sustainable Energy Technologies and Assessments*, 16:106–127. https://doi.org/10.1016/j.seta.2016.05.001.

Akinyele, D., Rayudu, R., and Nair, N. (2015). Development of photovoltaic power plant for remote residential applications: The socio-technical and economic perspectives. *Applied Energy*, 155:131–149. https://doi.org/10.1016/j.apenergy.2015.05.091.

Arnold, U. and Yildiz, O. (2015). Economic risk analysis of decentralized renewable energy infrastructures a monte carlo simulation approach. *Renewable Energy*, 77:227–239. https://doi.org/10.1016/j.renene.2014.11.059.

Baecker, P. N. (2007). *Real options and intellectual property: Capital budgeting under imperfect patent protection*. Springer Berlin Heidelberg.

Barrera, G. M., Ramrez, C. Z., and Gonzlez, J. M. G. (2016). Application of real options valuation for analysing the impact of public r&d financing on renewable energy projects: A companys perspective. *Renewable and Sustainable Energy Reviews*, 63:292–301. https://doi.org/10.1016/j.rser.2016.05.073.

Barros, J. J. C., Coira, M. L., la Cruz Lpez, M. P., and del Cao Gochi, A. (2017). Comparative analysis of direct employment generated by renewable and non-renewable power plants. *Energy*, 139:542–554. https://doi.org/10.1016/j.energy.2017.08.025.

Beaver, W. (1994). Nuclear nightmares in the philippines. *Journal of Business Ethics*, 13:271–279. https://doi.org/10.1007/BF00871673.

Bertsekas, D. P. (2012). *Dynamic Programming and Optimal Control*. Athena Scientific.

BNEF (2016). *New Energy Outlook 2016 Powering a Changing World*. Bloomberg New Energy Finance.

BNEF (2017a). *Global trends in clean energy investment*. Bloomberg New Energy Finance.

BNEF (2017b). *New energy outlook 2017*. Bloomberg New Energy Finance.

Brach, M. A. (2003). *Real Options in Practice*. Wiley.

Burtraw, D. and Krupnick, A. (2012). *The true cost of electric power: an inventory of methodologies to support future decision-making in comparing the cost and competitiveness of electricity generation technologies*. Renewable Energy Network Policy for the 21st Century (REN21).

Byrnes, L., Brown, C., Wagner, L., and Foster, J. (2016). Reviewing the viability of renewable energy in community electrification: The case of remote western australian communities. *Renewable and Sustainable Energy Reviews*, 59:470–481. https://doi.org/10.1016/j.rser.2015.12.273.

Cabalu, H., Koshy, P., Corong, E., E.Rodriguez, U.-P., and A.Endriga, B. (2015). Modelling the impact of energy policies on the philippine economy: Carbon tax, energy efficiency, and changes in the energy mix. *Economic Analysis and Policy*, 48:222–237. https://doi.org/10.1016/j.eap.2015.11.014.

Cardin, M.-A., Zhang, S., and Nuttall, W. J. (2017). Strategic real option and flexibility analysis for nuclear power plants considering uncertainty in electricity demand and public acceptance. *Energy Economics*, 64:226–237. https://doi.org/10.1016/j.eneco.2017.03.023.

Copeland, T. and Antikarov, V. (2003). *Real Options, Revised Edition: A Practitioners Guide*. W. W. Norton & Company, 1 edition.

Copiello, S., Gabrielli, L., and Bonifaci, P. (2017). Evaluation of energy retrofit in buildings under conditions of uncertainty: The prominence of the discount rate. *Energy*, 137:104–117. https://doi.org/10.1016/j.energy.2017.06.159.

Dai, H., Xie, X., Xie, Y., Liu, J., and Masui, T. (2016). Green growth: The economic impacts of large-scale renewable energy development in china. *Applied Energy*, 162:435–449. https://doi.org/10.1016/j.apenergy.2015.10.049.

Davis, G. A. and Cairns, R. D. (2012). Good timing: The econonomics of optimal stopping. *Journal of Dynamics and Control*, 36:255–265. https://doi.org/10.1016/j.jedc.2011.09.008.

Detert, N. and Kotani, K. (2013). Real options approach to renewable energy investments in mongolia. *Energy Policy*, 56:136–150. https://doi.org/10.1016/j.enpol.2012.12.003.

Dixit, A. K. and Pindyck, R. S. (1994). *Investment Under Uncertainty*. Princeton University Press.

DOE (2012). *Philippine Energy Plan 2012-2030*. Philippine Department of Energy.

DOE (2014). *Geothermal: cheapest energy source*. Philippine Department of Energy.

DOE (2016a). *2015 Philippine Power Statistics*. Philippine Department of Energy.

DOE (2016b). *Awarded Solar Grid*. Philippine Department of Energy.

DOE (2017). *2016 Philippine Power Statistics*. Philippine Department of Energy.

EEA (2010). *Estimated average EU external costs for electricity generation technologies in 2005*. European Environmental Agency.

EIA (2017a). *Annual Energy Outlook 2017*. Energy Information Administration.

EIA (2017b). *International Energy Outlook 2017 Table: World total energy consumption by region and fuel*. Energy Information Administration.

EIA (2017c). *Levelized Cost and Levelized Avoided Cost of New Generation Resources in the Annual Energy Outlook 2017*. Energy Information Administration.

Eissa, M. A. and Tian, B. (2017). Lobatto-milstein numerical method in application of uncertainty investment of solar power projects. *Energies*, 10:1–19. https://doi.org/10.3390/en10010043.

Emmanouilides, C. J. and Sgouromalli, T. (2013). Renewable energy sources in crete: Economic valuation results from a stated choice experiment. *Procedia Technology*, 8:406–415. https://doi.org/10.1016/j.protcy.2013.11.053.

Eryilmaz, D. and Homans, F. R. (2016). How does uncertainty in renewable energy policy affect decisions to invest in wind energy? *Electricity Journal*, 29:64–71. https://doi.org/10.1016/j.tej.2015.12.002.

Fagiani, R., Barquin, J., and Hakvoort, R. (2013). Risk-based assessment of the cost-efficiency and the effectivity of renewable energy support schemes: Certificate markets versus feed-in tariffs. *Energy Policy*, 55:648–661. https://doi.org/10.1016/j.enpol.2012.12.066.

Fernandez, L. (2015). *Power prices: Where we are and how do we reduce the bill.* Philippine Department of Trade and Industry (DTI).

Fleten, S.-E., Linnerud, K., Molnr, P., and Nygaard, M. T. (2016). Green electricity investment timing in practice: Real options or net present value? *Energy*, 116(1):498–506. https://doi.org/10.1016/j.energy.2016.09.114.

Fonseca, M. N., de Oliveira Pamplona, E., de Mello Valerio, V. E., Aquila, G., Rocha, L. C. S., and Junior, P. R. (2017). Oil price volatility: A real option valuation approach in an african oil field. *Journal of Petroleum Science and Engineering*, 150:297–304. https://doi.org/10.1016/j.petrol.2016.12.024.

FS-UNEP (2017). *Global Trends in Renewable Energy Investment 2017.* Frankfurt School - UNEP for Climate and Sustainable Energy Finance.

Gelabert, L., Labandeira, X., and Linares, P. (2011). An ex-post analysis of the effect of renewables and cogeneration on spanish electricity prices. *Energy Economics*, 33:S59–S65. https://doi.org/10.1016/j.eneco.2016.09.024.

Guedes, J. and Santos, P. (2016). Valuing an offshore oil exploration and production project through real options analysis. *Energy Economics*, 60:377–386. https://doi.org/10.1016/j.eneco.2016.09.024.

Gusano, D. G., Espegren, K., Lind, A., and Kirkengen, M. (2016). The role of the discount rates in energy systems optimisation models. *Renewable and Sustainable Energy Reviews*, 59:56–72. https://doi.org/10.1016/j.rser.2015.12.359.

Hach, D. and Spinler, S. (2016). Capacity payment impact on gas-fired generation investments under rising renewable feed-ina real options analysis. *Energy Economics*, 53:270–280. https://doi.org/10.1016/j.eneco.2014.04.022.

Hofert, M. and Wthrich, M. V. (2013). Statistical review of nuclear power accidents. *Asia-Pacific Journal of Risk and Insurance*, 7:1–18. https://doi.org/10.1515/2153-3792.1157.

Hong, G. W. and Abe, N. (2012). Sustainability assessment of renewable energy projects for off-grid rural electrification: The pangan-an island case in the philippines. *Renewable and Sustainable Energy Reviews*, 16:54–64. https://doi.org/10.1016/j.rser.2011.07.136.

IAEA (2016). *Nuclear power development program in the Philippines.* International Atomic Energy Agency.

IEA (2017). *Key world energy statistics.* International Energy Agency.

IEC (2016). *Regional/Global Comparison of Retail Electricity Tariffs.* International Energy Consultants.

Insley, M. (2002). A real options approach to the valuation of a forestry investment. *Journal of Environmental Economics and Management,* 44:471–492. https://doi.org/10.1006/jeem.2001.1209.

IRENA (2017). *Renewables Readiness Assessment: The Philippines.* International Renewable Energy Agency.

Jeon, C., Lee, J., and Shin, J. (2015). Optimal subsidy estimation method using system dynamics and the real option model: Photovoltaic technology case. *Applied Energy,* 142:33–43. https://doi.org/10.1016/j.apenergy.2014.12.067.

Jonsson, T., Pinson, P., and Madsen, H. (2010). On the market impact of wind energy forecasts. *Energy Economics,* 32:313–320. https://doi.org/10.1016/j.eneco.2009.10.018.

Junginger, M., Faaij, A., and Turkenburg, W. (2005). Global experience curves for wind farms. *Energy Policy,* 33:133–150. https://doi.org/10.1016/S0301-4215(03)00205-2.

Kaiser, J. C. (2012). Empirical risk analysis of severe reactor accidents in nuclear power plants after fukushima. *Science and Technology of Nuclear Installations,* 2012:1–6. https://doi.org/10.1155/2012/384987.

Kamjoo, A., Maheri, A., M.Dizqah, A., and A.Putrus, G. (2016). Multi-objective design under uncertainties of hybrid renewable energy system using nsga-ii and chance constrained programming. *International Journal of Electrical Power & Energy Systems,* 74:187–194. https://doi.org/10.1016/j.ijepes.2015.07.007.

Ketterer, J. C. (2014). The impact of wind power generation on the electricity price in germany. *Energy Economics,* 44:270–280. https://doi.org/10.1016/j.eneco.2014.04.003.

Kim, K., Park, H., and Kim, H. (2017). Real options analysis for renewable energy investment decisions in developing countries. *Renewable and Sustainable Energy Reviews,* 75:918–926. https://doi.org/10.1016/j.rser.2016.11.073.

Kim, K.-T., Lee, D.-J., and Park, S.-J. (2014). Real options analysis for renewable energy investment decisions in developing countries. *Renewable and Sustainable Energy Reviews,* 40:335–347. https://doi.org/10.1016/j.rser.2014.07.165.

Kitzing, L., Juul, N., Drud, M., and Boomsma, T. K. (2017). A real options approach to analyse wind energy investments under different support schemes. *Applied Energy,* 188:83–96. https://doi.org/10.1016/j.apenergy.2016.11.104.

Kumbaroglu, G., Madlener, R., and Demirel, M. (2008). A real options evaluation model for the diffusion prospects of new renewable power generation technologies. *Energy Economics,* 30:1882–1908. https://doi.org/10.1016/j.eneco.2006.10.009.

Lee, H., Park, T., Kim, B., Kim, K., and Kim, H. (2013). A real option-based model for promoting sustainable energy projects under the clean development mechanism. *Energy Policy*, 54:360–368. https://doi.org/10.1016/j.enpol.2012.11.050.

Lee, Y. S. F. and So, A. Y. (1999). *Asia's Environmental Movements: Comparative Perspective*. Taylor & Francis Inc.

Loncar, D., Milovanovic, I., Rakic, B., and Radjenovic, T. (2017). Compound real options valuation of renewable energy projects: The case of a wind farm in serbia. *Renewable and Sustainable Energy Reviews*, 75:354–367. https://doi.org/10.1016/j.rser.2016.11.001.

Meller, H. and Marquardt, J. (2013). *Renewable energy in the Philippines: Costly or competitive? Facts and explanations on the price of renewable energies for electricity production*. Deutsche Gesellschaft fr Internationale Zusammenarbeit (GIZ) GmbH.

Myers, S. C. (1977). Determinants of corporate borrowing. *Journal of Financial Economics*, 5:147–175. https://doi.org/10.1016/0304-405X(77)90015-0.

NEA (2015). *Projected cost of generating electricity*. Nuclear Energy Agency.

NEA (2016a). *Cost of decommissioning nuclear power plants*. Nuclear Energy Agency.

NEA (2016b). *Cost of major accidents: an introduction*. Nuclear Energy Agency.

OECD (2016). *OECD Business and Finance Outlook 2016*. OECD Publishing, Inc.

Paleco (2016). *Status of Electrification*. Palawan Electric Cooperative.

Pereira-Jr, A. O., da Costa, R. C., do Vale Costa, C., de Moraes Marreco, J., and Rovere, E. L. L. (2013). Perspectives for the expansion of new renewable energy sources in brazil. *Renewable and Sustainable Energy Reviews*, 23:49–59. https://doi.org/10.1016/j.rser.2013.02.020.

Pindyck, R. S. (1993). Investments of uncertain cost. *Journal of Financial Economics*, 34:53–76. https://doi.org/10.1016/0304-405X(93)90040-I.

Postali, F. A. and Picchetti, P. (2006). Geometric brownian motion and structural breaks in oilprices: A quantitative analysis. *Energy Economics*, 28:506–522. https://doi.org/10.1016/j.eneco.2006.02.011.

Pringles, R., Olsina, F., and Garces, F. (2015). Real option valuation of power transmission investments by stochastic simulation. *Energy Economics*, 47:215–226. https://doi.org/10.1016/j.eneco.2014.11.011.

Qiu, Y. and Anadon, L. D. (2012). The price of wind power in china during its expansion: Technology adoption, learning-by-doing, economies of scale, and manufacturing localization. *Energy Economics*, 34:772–785. https://doi.org/10.1016/j.eneco.2011.06.008.

Rangel, L. E. and Lvque, F. (2012). How fukushima dai-ichi core meltdown changed the probability of nuclear accidents? *Safety Science*, 64:90–98. https://doi.org/10.1016/j.ssci.2013.11.017.

REN21 (2016). *Renewables 2016 global status report*. Renewable Energy Network Policy for the 21st Century.

Ritzenhofen, I. and Spinler, S. (2016). Optimal design of feed-in-tariffs to stimulate renewable energy investments under regulatory uncertainty a real options analysis. *Energy Economics*, 53:76–89. https://doi.org/10.1016/j.eneco.2014.12.008.

Rothwell, G. (2006). A real options approach to evaluating new nuclear power plants. *The Energy Journal*, 27:37–53. https://doi.org/10.5547/ISSN0195-6574-EJ-Vol27-No1-3.

Santos, L., Soares, I., Mendes, C., and Ferreira, P. (2014). A real options approach to evaluating new nuclear power plants. *Renewable Energy*, 68:588–594. https://doi.org/10.1016/j.renene.2014.01.038.

Savino, M. M., Manzini, R., Selva, V. D., and Accorsi, R. (2017). A new model for environmental and economic evaluation of renewable energy systems: The case of wind turbines. *Applied Energy*, 189:739–752. https://doi.org/10.1016/j.apenergy.2016.11.124.

Shi, H. Y. and Song, H. T. (2013). Applying the real option approach on nuclear power project decision making. *Energy Procedia*, 39:193–198. https://doi.org/10.1016/j.egypro.2013.07.206.

Sisodia, G. S., Soares, I., and Ferreira, P. (2016). Modeling business risk: The effect of regulatory revision on renewable energy investment - the iberian case. *Renewable Energy*, 95:303–313. https://doi.org/10.1016/j.renene.2016.03.076.

Sovacool, B. K. (2010). A comparative analysis of renewable electricity support mechanisms for southeast asia. *Energy*, 35:1779–1793. https://doi.org/10.1016/j.energy.2009.12.030.

Stich, J. and Hamacher, T. (2016). The cost-effectiveness of power generation from geothermal potentials in indonesia and the philippines. *IEEE Innovative Smart Grid Technologies - Asia*, 2016:177–182. https://doi.org/10.1109/ISGT-Asia.2016.7796382.

Tian, L., Pan, J., Du, R., Li, W., Zhen, Z., and Qibing, G. (2017). The valuation of photovoltaic power generation under carbon market linkage based on real options. *Applied Energy*, 201:354–362. https://doi.org/10.1016/j.apenergy.2016.12.092.

Tian, L., Shan, H., and Zhu, N. (2016). Analysis of the real options in nuclear investment under the dynamic influence of carbon market. *Energy Procedia*, 104:299–304. https://doi.org/10.1016/j.egypro.2016.12.051.

Tietjen, O., Pahlea, M., and Fuss, S. (2016). Investment risks in power generation: A comparison of fossil fuel and renewable energy dominated markets. *Energy Economics*, 58:174–185. https://doi.org/10.1016/j.eneco.2016.07.00.

Trigeorgis, L. (1996). *Real Options: Managerial Flexibility and Strategy in Resource Allocation*. MIT Press.

Twidell, J. and Weir, T. (2015). *Renewable Energy Resources*. Routledge, 3 edition.

Utama, N. A., Ishihara, K. N., and Tezuka, T. (2012). Power generation optimization in asean by 2030. *Energy and Power Engineering*, 4:226–232. https://doi.org/10.4236/epe.2012.44031.

Viallet, C. and Hawawini, G. (2015). *Finance for Executives: Managing for Value Creation*. Cengage Learning Emea, 5th edition.

Wang, X. and Du, L. (2016). Study on carbon capture and storage (ccs) investment decision-making based on real options for china's coal-fired power plants. *Journal of Cleaner Production*, 112:4123–4131. https://doi.org/10.1016/j.jclepro.2015.07.112.

Weibel, S. and Madlener, R. (2015). Cost-effective design of ringwall storage hybrid power plants: A real options analysis. *Energy Conversion and Management*, 103:871–885. https://doi.org/10.1016/j.enconman.2015.06.043.

Wesseh-Jr., P. K. and Lin, B. (2015). Renewable energy technologies as beacon of cleaner production: a real options valuation analysis for liberia. *Journal of Cleaner Production*, 90:300–310. https://doi.org/10.1016/j.jclepro.2014.11.062.

Wesseh-Jr., P. K. and Lin, B. (2016). A real options valuation of chinese wind energy technologies for power generation: do benefits from the feed-in tariffs outweigh costs? *Journal of Cleaner Production*, 112:1591–1599. https://doi.org/10.1016/j.jclepro.2015.04.083.

Xian, H., Colson, G., Mei, B., and Wetzstein, M. E. (2015). Co-firing coal with wood pellets for u.s. electricity generation: A real options analysis. *Energy Policy*, 81:106–116. https://doi.org/10.1016/j.enpol.2015.02.026.

Yang, M., Blyth, W., Bradley, R., Bunn, D., Clarke, C., and Wilson, T. (2008). Evaluating the power investment options with uncertainty in climate policy. *Energy Economics*, 30:1933–1950. https://doi.org/10.1016/j.eneco.2007.06.004.

Yeo, K. and Qiu, F. (2003). The value of management flexibilitya real option approach to investment evaluation. *International Journal of Project Management*, 21(4):Pages 243–250. https://doi.org/10.1016/S0263-7863(02)00025-X.

Zhang, M., Zhou, D., Zhou, P., and Chen, H. (2017). Optimal design of subsidy to stimulate renewable energy investments: The case of china. *Renewable and Sustainable Energy Reviews*, 71:873–883. https://doi.org/10.1016/j.rser.2016.12.115.

Zhang, M., Zhou, P., and Zhou, D. (2016). A real options model for renewable energy investment with application to solar photovoltaic power generation in china. *Energy Economics*, 59:213–226. https://doi.org/10.1016/j.eneco.2016.07.028.

Zhu, L. (2012). A simulation based real options approach for the investment evaluation of nuclear power. *Computers & Industrial Engineering*, 63(3):585–593. https://doi.org/10.1016/j.cie.2012.02.012.

Bisher im Logos Verlag Berlin erschienene Bände der Reihe

UA Ruhr Studies on Development and Global Governance

ISSN: 2363-8869

Vormals "UAMR Studies on Development and Global Governance"
(ISSN: 2194-167X, Vol. 61-64) und
"Bochum Studies in International Development" (ISSN: 1869-084X, Vol. 56-60)

Alle erschienenen Bücher können unter der angegebenen ISBN-Nummer direkt online
(http://www.logos-verlag.de) oder per Fax (030 - 42 85 10 92) beim Logos Verlag
Berlin bestellt werden.